MAINLINE FARMING
FOR CENTURY 21

Mainline Farming for Century 21

Acres U.S.A.
P.O. Box 91299
Austin, Texas 78709 U.S.A.
(512) 892-4400 • fax (512) 892-4448
info@acresusa.com • www.acresusa.com

Printed in the United States of America

Publisher's Cataloging-in-Publication

Skow, Dan.
Mainline farming for century 21 / Dan Skow, Charles Walters; — Austin, TX, ACRES U.S.A., 1991
    xvi, 208 pp., ill., 23 cm.
    Includes Index
    Includes Bibliography
    ISBN 0-911311-27-0 (trade)

    1. Sustainable agriculture. 2. Plants — nutrition. 3. Agriculture — United States. I. Skow, Dan. II. Walters, Charles. III. Title.

S591 .S57 1991   630.2

Dedicated to the memory of
Carey A. Reams,
with the expectation that his brand
of agronomy will sweep the nation
going into Century 21.

THE RULES OF CAREY A. REAMS

1. Carbon is the governor of moisture. One part carbon will hold four parts water.

2. The more carbon in a seed, the quicker it will sprout.

3. Manganese is the element of life. It brings the electrical charge into the seed.

4. All elements in a molecular structure are the same size under the same temperature and pressure.

5. The center core of an element tells whether it is an anion or a cation.

6. Nature will follow the line of least resistance.

7. The greater the density of the soil without humus, the greater the specific gravity of the soil.

8. The lesser the density of soil nutrients, the smaller the yields.

9. The greater the density of soil nutrients, the greater the yields.

10. The process of osmosis is not limited by time.

11. The less time it takes to grow something, the better the quality.

12. The higher the sugar and mineral content of plants and trees, the lower the freezing point.

13. Top quality produce will not rot, it will dehydrate.

14. All organic fertilizers are cationic.

15. Plants live off the loss of energy from the elements during the synchronization of these elements in the soil.

16. See everything you look at.

17. Like things attract each other.

18. For every cause there is an effect.

19. Phosphate controls the sugar content of a product.

20. The higher the phosphate content of a soil, the higher the sugar content of the crop. The higher the sugar content, the higher the mineral content. The higher the mineral content, the greater the specific gravity of a given bushel, box, bale, etc. The greater the specific gravity of the product, the healthier the animal.

21. All elements, except nitrogen, go into the plant in the phosphate form.

22. The ratio of all crops (except grasses) for phosphate and potassium in the soil is two parts phosphate to one part potassium. [2 P_2O_5 to 1 K.]

23. The ratio for all grasses is four parts phosphate to one part potassium. [4 P_2O_5 to 1 K.]

24. Potash determines the caliber of the stalk and leaves, the size of the fruit, and the number of the fruit which set on the trees.

25. Nitrogen is the major electrolyte in the soil.

26. Nitrogen is the sun in every molecule.

27. Only that plant food which is soluble in water is available to the plant.

28. Cationic substances go down.

29. Anionic substances go up.

CONTENTS

A NOTE FROM THE PUBLISHER

A few years ago, when we published *An Acres U.S.A. Primer*, it seemed appropriate to include as a glossary entry a definition of the *scientific system*. We concluded that "Most of what is generally called the *scientific system* is not science at all, but merely a procedural aspect that calls for setting up experiments that eliminate other possibilities, or it deals with making instruments that enable the investigator to find what he is looking for. The backbone of the scientific system has to do with asking the right questions. A scientist can only ask the right questions after his life has absorbed the experiences that lead him to a vision of the Creator's handiwork, hence the right question. In the final analysis, new discovery is accomplished by the mind and soul of the whole person and cannot be a mechanical scientific procedure. It stands to reason that you can't get the answers if you don't know the questions. When science falters, it is because no one is asking the right questions."

Few people have asked the right questions with the total consistency of Dr. Dan Skow. His teacher was the late Dr. Carey Reams, and the lessons Skow learned — and improved upon — were strong meat. They led him to a full appreciation of cause and effect, the role of precedence — every phenomenon or act preceding another — called *cause*. In terms of experimentation, cause is little more than an order of succession in a time frame. As one cause triggers an effect, responsibility fades away, and yet it never disappears entirely before reaching back to the origin of everything, to the creation — to philosophy's First Cause.

One additional note might clarify the wonderful odyssey of Dan Skow as depicted in this little volume. The scientific system cannot function without addivity, namely units of measurement that have equal intervals and an absolute zero. As we consider these ideas, history soon enough enters the picture, and so does one of man's first intellectual pursuits. This business of units for measurements goes back to the cradle of agriculture and a symbol one finds ten miles west of the modern city of Cairo. At the end of an acacia, tamarind and eucalyptus lined avenue lies the rocky plateau of Giza. It is a mile square, and from a height of 130 feet it dominates the palm groves of the Nile Valley. On this man-leveled plateau stands the great pyramid of Cheops. West lies the Libyan desert.

The great pyramid covers thirteen acres, or about seven midtown blocks in the city of New York. From this huge area, leveled to within a fraction of an inch, more than two and a half million limestone and granite blocks rise in 201 stepped tiers to a height of forty stories. These blocks weigh from two to 70 tons, the biggest being as heavy as a modern locomotive engine.

Time has taken most of the records associated with the enterprise, but we do know that whoever built the Great Pyramid knew the dimensions of this planet as they were not known again until the seventeenth century. By observing the stars, these builders could and did measure the day, the year, the hour, the minute, the second. They linked time with distance because they knew a circle had 360 degrees and accordingly they computed longitude and latitude, the distance from pole to pole, from earth to sun, and the exact equator. They graduated distances into exact units called cubit and foot, and no man could tamper with these measures because they were fixed by the vault of the heavens and recorded in the dimensions of the Great Pyramid.

In mathematics these ancients were advanced enough to have discovered the function of pi and Fibonacci's series. The record they left in deathless stone reveals that Eratosthenes was not the first to measure the circumference of the earth, Hipparchus was not the first to understand trigonometry, Pythagoras was not the first to originate his theorem, and Mercator was not the first to invent that projection.

The ancients also knew about dilution and delivering a uniform nutrient fix to their delta acres. One marvels at how nearly they approximated Dan Skow's distribution of atoms over acres, having first harvested carbon and paramagnetic stone from the Blue and White Nile.

This proximity of ancient Egypt to the lessons of the Bible is not lost in Skow's farming testament. Time after time Skow — standing on the shoulders of Carey Reams, Albert Einstein and Frank LeMotte — amplifies and clarifies what he has to say with biblical notes. The reader will be able to judge how well Skow's exposition and analysis of *cause* proves and upholds the *scientific system*.

Dan Skow was awarded suitable credentials from veterinary school in Ames, Iowa in 1968, after which he started to practice animal medicine. It wasn't long before he realized that, indeed, it was practice, and the time had come for him to get a serious education. A few years later he heard about Carey Reams in Florida. Reams had backgrounded a little book called *The Curse Causeless Shall Not Come*, which was promptly proscribed by certain federal agencies which sent the journalist who wrote it to jail.

Dan Skow sought out Carey Reams. As a veterinary practitioner, he'd come to realize that he couldn't take a situation by the nape of the neck and the seat of the pants, and shake out a result without making the scene. Reams didn't turn out to be the greatest orator in the world. Nor was he the greatest writer, but his ideas merited being weighed out on a jeweler's scale. Over the next ten or twelve years, Skow learned everything he could from Carey Reams, exhibiting the patience of Job in the process.

He also learned about the triumvirate named above — Reams, Einstein and LeMotte — and the fine-tuned system of health and soil management their insight, measurements and vision had accounted for. Reams had corresponded and counseled with William A. Albrecht of the University of Missouri, whose *Albrecht Papers* we now keep in print. By the time Skow met Reams, he remained a lone hetman, and listeners were few. Almost alone in the agricultural field, Dan Skow listened.

The education of Dan Skow did not stop with the lessons and wisdom of Carey Reams, for this veterinarian went on to account for developments well beyond the range of knowledge he had learned from his predecessors.

One thing above all else caught the attention of the young practitioner. When Carey Reams opened a seminar, morning devotions led the way. "I used to sit there, " Dan Skow recalled after Reams had passed from the scene, finally a victim of the war wounds he received in the Philippines, "and I was at least as intrigued with what he had to say about the Scriptures and the life of Jesus Christ as I was about soil programs. Once I came to comprehend that opening part of his lecture, the door flew open and I understood the rest."

Speaking before an *Acres U.S.A.* meeting in 1989, Skow confided that he had learned a lot of interesting things in college. He'd learned, for instance, what he needed to put on a piece of paper to get a grade, but these things meant little to him then, and not much more now. He recommended an *Acres U.S.A.* article about the Luebke family in Austria, which appeared in the December 1989 issue. "It dispels many of the myths that are being taught," he said, and then he proceeded to dispel a few more on his own.

Dan Skow deals in plain mathematics. This means the abstractions of numbers are involved, but if you learn the lessons well you'll soon realize that when a plant is sick or not growing right, a different math is involved.

Mainline Farming for Century 21, by Dan Skow and Charles Walters, is a graduate course in agriculture. It does not settle for tip-toeing beyond the so-called conventional, it takes bold strides — and it answers what the farmer wants to know, all the while demolishing the mythology on which toxic rescue technology has been built.

FOREWORD

The "I" in the text that follows is Dan Skow, D.V.M., the most important of the students now standing on the shoulders of the giant known as Dr. Carey Reams. The task of setting down what Dr. Skow learned during a decade of tutelage fell to *Acres U.S.A.* editor and publisher Charles Walters, who added peripheral research to the effort.

The end product has to be considered a joint effort by these three people. As with many end products, *Mainline Farming for Century 21* is not an end product after all. It is merely an informed probe for the purpose of setting up some of the right questions. Its objective is to analyze the terrain on which we stand, and the steps that must be taken to deliver agriculture out of the thrall of toxic technology into Century 21 as a production engine capable of delivering high quality uncontaminated food without foisting on the environment land and water pollution.

Some few esoteric rules and a great deal of folklore will be left behind by this process, the folklore of toxic technology being first in line for its scheduled demise.

Withal, *Mainline Farming for Century 21* is a book for our times. It asks no quarter and stands ready to defend itself against all reasoned assaults. Yet it represents no more than a beginning, an exercise in asking the right questions and explaining the right answers.

1

THE PUBLIC LIFE OF A CELL

It is both a single cell and a magnet, this beautiful planet on which we live. It is positioned in a bathhouse of energy, and at ground level it drifts apart and reforms itself on crustal plates afloat on a core of fire. It is perpetually in motion. The greatest and unlooked-for bonus of our space program — a color picture of the planet itself — suggests the earth is a living organism tied by a gravitational umbilical cord to its nebula, the sun. Yet this may be an incorrect assumption. The earth really is more like a cell than an organism. It is self contained and superbly skilled at handling its source of energy, the sun. The earth gives us both an overview and a common denominator, for in dealing with agriculture we are forced to consider — as far as we are able — the cell and the atoms of elements in order to discern the Creator's handiwork.

Let's consider the cell ever so briefly before we paint, as broad brush strokes, the role of uncommon good sense in farm production.

The cell — platelike, elongated, concave, disclike, spherical — is both a fiction and a truth. It fits into everything that is biologically alive, bacteria or fungus, plant, animal or human being. It can't hold any of the shapes assigned to it by artists because of close packing, and yet it provides us with a bird's eye view of Creation. This view must always remain in focus, whether the farmer is planning a fertility program for his crops, or the nutritional regimen for animal protein production, or a program for human health. In designing a program to grow any living thing, the *open sesame* for success is full understanding and measuring the atoms of the elements. Nature permits no exception, and this sets up the importance of nitrogen. The smallest bud and the largest branch are dependent on nitrogen, otherwise the cell won't grow. To make this statement is not to indulge in single-factor analysis. We know and imply that all energy comes from the sun. Plants capture solar energy in their leaves, and use this energy to make sugar. They go on to construct cells of all kinds, using sugar, air, water and earth minerals as building blocks. The first part of this equation is called photosynthesis, which means plants use sunlight to synthesize a new substance, namely sugar. After that, sugar and oxygen serve up carbon dioxide, water and energy.

This unfolding sequence does not happen by accident. The Creator has evolved gene strips to help govern the life process. All life forms on earth have these inherited gene strips. They determine cell structures, stamina, size, shape, color, odor — on and on.

Plants also have enzymes. These are small protein units that act as on-scene engineers in the cell building business. They take raw materials, such as earth minerals, and see to it that they reach the right stem, root, bud, flavor, or whatever. Indeed, how enzymes create hot spots to attract essential cell building materials — iron, nitrogen, boron, for instance — so that they can be linked to the right molecules in plant cells must be considered a miracle. Equally a miracle is the fact that most farm crops are 95% sunshine, air and water, and only 5% earth minerals.

It should be at once apparent that single-factor analysis is a mark of the amateur. And yet we are all obliged to find our common denominators and measure our equations according to the law of the little-bit— the minimum, if you will. Many forms of nitrogen figure in this cell building business, and all of them must be understood. For now it will be enough to understand basic nitrogen itself. Without it, plants, animals and human beings will not grow. Indeed, a nitrogen shortfall will preside over the deterioration of health — plant, animal or human — with economic performance of the first and the second fading away.

Protein contains nitrogen and nitrogen is essential to the construction of a new cell. That statement seems simple enough. It leads logically to the syllogism that says deficiency short-circuits the build-up of a cell and signals the beginning of disease. In human beings this wasting away of health is called degenerative metabolic disease, and for plant life a whole nomenclature of Latinized descriptions has been invented, most of it describing a decay and absence of healthy growth. Incredibly, the *1959 Yearbook of Agriculture* blandly proclaimed that "Lack of fertilizer may reduce the yield of a crop, but not the amount of nutrients in the food produced." It is the continued iteration and reiteration of this nonsense that prompts me to set down both what I have learned from Carey Reams, and what continued investigations and practice have revealed. I find it redundant — and still necessary — to note that calcium, nitrogen, phosphorus and a long list of elements, including the trace elements, are required to synthesize amino acids, proteins, vitamins, enzymes, lipids, octocosanols, phosphatides and the rest of the building blocks used to construct plant life. Plants and microbes — even those in the cow's gut — synthesize the amino acids that make up proteins. As William A. Albrecht put it, "Both plants and animals assemble their proteins to provide their reproductive functions, since these are the only compounds through which the stream of life can flow."

The childish belief that this complicated life process can be serviced with simplistic N, P and K fertilization and anti-nature farm management should suggest that man is, indeed, happy in ignorance.

Water is essential. I have been called to farms with sick and dying animals only to find missing the most essential ingredient,

GRAMMAR OF THE SUBJECT

It has been noted often that it is difficult to read material about various forms of farming without understanding the grammar of the subject. The late Carey Reams used grammar that was sometimes at a variance with definitions handed out as classroom fare in school. Carey Reams was clear and concise at all times. He has been unjustly criticized for not having a bibliography to defend, and for not chiseling his words into stone called "the literature." The critics must remember that during most of the years Carey Reams was active, it was literally a crime to take issue with the conventional science. People like Wilhelm Reich went to jail for "writing it down." So did Victor Irons, the discoverer of Green Life. Carey Reams also endured the flight from reason that held sway for most of his active years. Only in the twilight of his career did he write anything down for his most serious student, Dan Skow, D.V.M. We are all standing on the shoulders of giants when we distribute knowledge gained at the feet of the masters. Carey Reams was such a master, as the arithmetic in this chapter readily illustrates.

just plain water. And yet water is vital to life of any kind. Within a biological system, water can be broken down into hydrogen and oxygen, and these two elements control the life equation. They transport nutrients through cell walls. They govern too many other "single factors" to list in a short text.

I get calls from farmers who want to grow 200 bushels of corn. I admit I have to chortle before I ask, *And how much water do you have available?* It makes little difference how much plant food is introduced into the rootbed if there is no water. You have to look at the basics — essential things first — and then go on from there. I've had farmers call me to ask about what kind of biologicals to put on, and each time I have had to ask about the availability of water.

The next essential in this growth equation is carbon. Most plants get their basic carbon from the carbon dioxide in the air, which brings us back to carbon and oxygen again. Nitrogen, hydrogen, oxygen and carbon all come out of the air. They are nature's gifts of nutrients, and it is a sad thing to observe that most farmers do not know how to accept a free gift. One and all, growers seem to have the idea that they must do something for these gifts.

The things farmers do to avoid accepting nature's gifts would be ruled from fiction if a novelist had to invent them. Rather than accept free nitrogen, they gas the crop to death with anhydrous ammonia. *National Geographic* carried an article in the February 1990 issue which told about the Aral Sea, a body of water greater than any of our Great Lakes, save Superior. This used to be a gigantic sea in Central Asia. The article told how chemical agriculture had swept that part of the world, how defoliants for cotton turned miles and miles of land into desert. They diverted — for irrigation — the Amu Darya and the Syr Darya rivers that emptied into the Aral Sea. As the sea retreated, dust storms — chemical storms, actually — swept the land, leaving in their wake the scourge of hepatitis. The incidence of cancer is reported at 50%, and few children live to the age of ten. There is a sign in the city of Nukus that reads, THE ARAL WILL LIVE AGAIN. No one believes it. A more contrite legend was chalked on a fishing boat, now high and dry. It reads, FORGIVE US ARAL. PLEASE COME BACK. This is what is happening in the Soviet Union, and I can see it happening up the road in Iowa.

I have seen farmers grow alfalfa, then cut it and watch dehydration virtually make it evaporate. I mention this to stress again why a farmer needs to understand how a cell is made. When you have a problem with watery crops, calcium is missing in that cell. Calcium is the king of nutrients, and yet it rates little mention when the bag of fertilizer is discussed. I recommend several books to serious growers — *An Acres U.S.A. Primer, The Anatomy of Life and Energy in Agriculture, The Albrecht Papers.* These are must reading. There is one title, *Fertilizers and Soil Amendments*, which has more sound discussion of calcium than any other book I know of. It tells about the importance of calcium in plant production, and it explains the role of adequate calcium in the cell for maintenance of plant health.

Unlike nitrogen, oxygen, hydrogen and carbon, calcium does not come from the air. It has to come from the soil. Calcium in the soil is very insoluble. It has to be acted upon by organic acids which are produced by plant roots, bacteria, yeasts and fungi in the soil. Without this activity, calcium cannot be incorporated into the plant structure.

I have heard this story many times. A farmer puts on a ton of lime and he sees no response. He fails to correct other factors. He fails to account for water and carbon — which is plant residue — so there is nothing for bacteria to consume to sustain the process by which calcium is made available. These things are so basic, I confess I am at a loss to understand how they can be overlooked.

Next, I am told that I am a veterinarian — which I am — and that I shouldn't know anything about soil systems and agronomy. Actually, there are more animals in the soil than there are in the piglot or cow feedlot. I figured out how to feed swine and cattle and a few other animals, and more recently I found out that bacteria need the same things. These little critters need water. The number-one essential element for all micro-creatures is calcium.

We talk about oxygen. *I have to wonder about farmers who drive over their fields, compact the soil, and squeeze out all the* oxygen or carbon dioxide — and then fail to remember that they need calcium to give quality to cell structure and to make the perimeter of the cell strong and firm. Those who serve farmers as advisors have lost track of the fact that alfalfa, for instance, is supposed to be solid stemmed. Obviously, solid stemmed alfalfa will stand up. It may sway in the wind and rain, but it won't lay down. Small grains — barley, wheat, oats, spelts — all should have solid stems. When Carey Reams first told me that, I nearly bowed out. I figured he'd gone a little too far.

Some years ago, the late William A. Albrecht made a talk which was later published, and is kept available by Acres U.S.A. as an audio tape. He called it Diagnosis and Post-Mortems and it was his point that we no longer know what healthy organs in animals look like. Certainly we no longer remember what healthy small grain stems should look like. I had a farmer call me when he got his first wheat with solid stems. He was so excited he stopped his combine when he saw that stubble with its pearly white color. The yield was exceptional. And yet for a moment this fellow figured he'd run into a new disease.

How many farmers really know what corn is supposed to look like? On the left is the kind of corn most farmers raise.

And on the right is the same corn kernel from the side view. When you have such a kernel, you have lost 50% of your production. This is an anemic, sick kernel of corn. It should have a nice rounded top, and exhibit square lines with only a minor point at the top.

On the following page is a healthy kernel of corn, fully filled, and without the commonly seen dent. An end view of the same kernal illustrates the health and maturity of the corn. Such a grain will field dry and deliver the best possible weight and nutrient load per bin or bushel.

In any pecking order of elements that I use as a preliminary introduction to the grammar of the subject, I have to list phosphate. Phosphate is basically a catalyst. In the process of photosynthesis, it takes phosphate to combine carbon dioxide from the air and water up through the roots to form sugars. Sugars are nothing more than carbon, hydrogen and oxygen. The wrench or crowbar to opening the door to good sugar and nutrient and energy levels in plants is having plenty of available phosphate. It works as a catalyst. But it may as well be understood up front that you can never achieve the super crops we have harvested — and *Acres U.S.A.* has written about — by using acid treated phosphate. It simply will not happen. It may be possible to make some improvements, but you will never achieve top quality crops. There will be no nice and rounded corn kernels without dents. There will be no solid stems in small grains and alfalfa. Not even adequate water, oxygen, carbon dioxide and nitrogen will serve to deliver quality crops without phosphate in the right form.

There are certain fertility elements that account for foliage, leaves, flowers and little else. I can grow corn and get only a stalk, but no ear. I can grow tomatoes with picture pretty plants, but no fruit. In fact, I know a farmer in Pennsylvania who did just that. He grew some thirty acres of tomatoes under contract for a baby food company. In the middle of the summer he had the most beautiful vines, albeit no tomatoes. Not one of the field men or college experts could tell him what to do.

The best and basic sources — if you know how to manage them — are soft rock phosphate and hard rock phosphate. If hard rock phosphate is used, a hyper-active bacterial system is a must.

Bacterial action is a must in any case when working with either hard or soft rock.

Phosphoric acid works well in foliar sprays. Generally speaking, acid-treated phosphates should be relegated to crutch status. Monammonium phosphate — known in trade channels as 11-50-0 — will never permit the best buildup of sugars in plants, but it may be defended as a crutch to get cash flow while better procedures are started.

Potassium will also do many similar things, such as growing foliage. But there is a problem. When an excess of potassium is applied, it replaces calcium and launches disease. Too often farmers apply still more potassium to cure the problem only to get still more disease. If potassium is replacing calcium in the leaf, both the leaf and the stem will exhibit small black specks. This is often diagnosed as blight, but treatment with sprays won't make the spots disappear. Potassium is essential for growth, but it is easy to fertilize with too much. Potassium in soil is fairly soluble. Calcium is fairly insoluble. Nature has ordered microorganisms into the soil to manage the ratios. But when chemicals of organic synthesis annihilate that valuable livestock in the soil, plants substitute potassium for calcium, always exacerbating disease problems, always setting up the ultimate embarrassment. Cows go down on bad feed. The classic signs and symptoms are bad kidneys. Hogs become arthritic. Dairy animals get mastitis and somatic cell counts go through the roof.

Some farm crops go directly to the dinner table. In crops where the calcium has been replaced by potassium — lettuce, broccoli, brussels sprouts, spinach — this potassium-calcium imbalance causes heart trouble and kidney disease. Using the conventional N, P and K fertilization program, agronomy puts too much potassium in the system, and not enough organically soluble calcium.

Chlorides also account for cosmetic growth, which may or may not explain the enchantment many growers have with potassium chloride. It works, but *works* has to be interpreted loosely. The response is both obvious and temporary — and costly in the long run.

PHOSPHATE-POTASSIUM RATIO

Potassium determines three basic things in plant growth: the thickness of the leaf, the thickness of the stem, and the caliber of a corn stalk or alfalfa stem; it determines the number of fruit that sets on a plant . . . it is the binder that holds the fruit to the stem; it determines the size of the fruit.

The ratio of phosphate to potassium in the soil should be 2:1–two parts phosphate to one part potash. This means that for maximum yields a minimum of 400 pounds of phosphate and 200 pounds of potash is indicated. This ratio and level applies to all crops except grasses. On grasses a ratio of four parts phosphate and one part potash is correct. Alfalfa has the ability to take practically all its potash from the air. Therefore, it needs very little from the soil.

Potash can be obtained from many things. Some good sources are sulfate of potash, Chilean nitrate of potash, hardwood ashes, tobacco stems, pecan hulls, sawdust, wheat or oat straw, and chicken manure. Sawdust is about the best source of potash. It also serves as a nutrient for bacteria. It contains carbon, which causes the soil to hold more moisture, and also has many trace elements. If you need potash quickly, 100 to 200 pounds of sulfate of potash per acre will give you some almost immediately. Sawdust and the other organics take about 90 days to become available.

Muriate of potash is one fertilizer that ought to be completely banned. It contains 40 to 50% chlorine and is actually potassium chloride. It takes only two parts per million of chloride in water to completely kill all bacteria. So 200 pounds per acre of muriate of potash is fifty times more chloride than it takes to kill all bacteria.

Trees and plants have a prenatal period just as do animals. In the North Temperate Zone, from July 20 to September 15, trees take in potassium in the form of Sul-Po-Mag (sulfate of potash magnesia). An application of 200 pounds per acre can be made at this time, and is usually sufficient for eight to ten years.

The first thing you will notice in a crop with an excess of potash is that the leaves will turn black and die at the tip. The second thing is that little black spots appear along the sides of a leaf. Little black dots on the stem of an alfalfa plant also indicate an excess of potash.

The other element that governs basic growth is nitrate nitrogen. You can really stimulate lush growth with this nitrogen form, but cattle turned out on such feed announce their continued hunger with memorable sounds. Plant foods that cause seed production are ammonial nitrogen, phosphorus, metal trace nutrients, manures and composts. It is this nitrogen requirement that brings us back to a hard look at the cell. Cells are not merely a neat packet of life's parts. They are rented and occupied by the mitochondria, which engineer the oxidative energy used in the "living" core. They replicate themselves, taking instructions from their own DNA and RNA. Much like rhizobial bacteria on the roots of beans, they are symbionts, and they govern an ecosystem as complicated as the great earth cell itself.

The American fertilizer industry has made life difficult for a serious grower. Aqua ammonia, for instance, is unavailable. In the case of the tomato grower with all leaves and vines and no fruit, I told him to buy Bo-Peep ammonia at the local grocery store, enough to apply one quart per acre. The cell needed this form of nitrogen — and it needed water as well. So I told the tomato grower to purchase apple cider vinegar for carbon. A half gallon of vinegar with a quart of Bo-Peep in 20 gallons of water made a perfect spray for the crop. Forty-eight hours later that tomato patch sported the most beautiful layer of blossoms ever. At the end of a week, the tomato patch was loaded with marble-sized tomatoes. A week later some of the fruit started to drop off. After a second spraying the crop took off again. We did that four times during the season. Later in the season I tripped to Lancaster, Pennsylvania to see the results. That farmer couldn't get the entire crop harvested.

The bottom line here is that there are nutrients that force a plant into seed production, and there are those that provide for growth. Once the seeds are started, the emphasis shifts from seed production back to growth.

By weight and volume, calcium is needed more than any other element. Some of the schoolmen types object that great calcium overloads do not show up via plant analysis. But they fail to understand that calcium is essential *for its energy creation potential in the soil to release the other elements that cause a plant to grow.* It isn't needed necessarily in the crop itself. That is why high calcium lime is indicated, one with no more than 5% magnesium evident in its test content. A good calcium source is calcium sulfate, better known as the compound gypsum. Calcium nitrate is an excellent source of water soluble calcium for spray application or touch up work. It is usually too expensive to be a whole source of calcium. Bone meal is excellent, but not economically sound for large-scale farming situations. There are other sources that won't generate trouble for the grower.

Crops that need a lot of calcium are alfalfa — unless you're going to harvest the crop for seeds — lettuce, cabbage, broccoli, brussel sprouts and spinach. If you want really crisp lettuce, calcium confers that crispness to the outer cell wall.

In making correct decisions, certain tests are indicated. The anatomy of these tests will be deferred to a later section in this book. For now it is enough to say we have to get into the ball park. It is also enough to remember that the only time you have ammonia in a test is when active bacteria in the soil are dying off and storing their nitrogen from season to season.

Most American farmers will never discover this key. It is common enough to find plenty of nitrate nitrogen on an initial test, but most soils do not have enough humus to hold the right levels. Phosphates and potassium oxide should be available in a 2:1 ratio for seed production crops. Calcium and magnesium should be about 7:1. Most farmers have a 3:1 or even a 1:1 ratio. Any ratio narrower than 5:1 means problems beyond instant comprehension. It means compacted soils, bacteria that can't proliferate, and weed takeover — in short, a marginal production sequence. For every pound of water-soluble magnesium in the soil, one pound of nitrogen is released straight into the air. This means that until you get the ratio correct, you are going to have to add increasing amounts of nitrogen to grow a crop that will support payment of bills.

The few observations set down here merely hint at the complexities that rule the crop production sequence. In the chapters that follow we will examine and explain the Reams Biological Theory of Ionization. In doing so, we will re-examine some of the most cherished ideas protected by schoolmen and chemical technology. We will discern that disease is not some kind of demonology, and pathogenicity is not even the rule. We will learn about an agronomy that is never panic stricken. We will see the intelligent farmer for what he is, a human being fired by energy from microbial symbionts lodged in its cells, governed by codes of nucleic acid, educated and informed by neurons, living off free capital from the sun, and now in charge of the farm, for better or for worse. We have our mitochondria, plants have their chloroplasts, both permanent residents. If we are successful in this endeavor, we will both feel our conjoined intelligence operative in the production of better crops and animals.

2

ATOMS PER ACRE

An American acre has 43,560 square feet, or 4,840 square yards. It provides us with the first requirement for scientific farming, namely a suitable and accurate unit of measurement. There are other "acres," although they are rarely used. The Scottish acre contains 6,150 square yards; the Irish acre, 7,840 square yards; the Cheshire acre, 10,240 square yards. But the acre defined for us by the Weights and Measures Act of 1878 gave us our 43,560 square feet unit — one so suitable it is preferred around the world, even where hectares and metric standards are official public policy.

Thomas Jefferson rejected the metric system because it failed to keep faith with the precept of weight, distance and time coordinated with the vault of the heavens. Thus the Weights and Measures Act of 1878 as written, and thus our own acres connection when we compute plant nutrients, atom by atom, under the guidance of the Biological Theory of Ionization.

The Biological Theory of Ionization takes ions — cations and an-ions — apart and puts them back together, so to speak. Some plants seem to grow in obedience to the Fibonacci number series

1, 2, 3, 5, 8, 13 . . .

and so on, cell growth being governed by the two previous divisions, and not in geometric progression, the last total being the sum of the previous two. Some plants seem to grow accordingly to the square of numbers rule, others follow more complicated formulas. We do not know why. There are many things we have to accept as the Creator's plan. Sometimes we can discover cause and effect a few causes deep. Often we have to describe the phenomena and proceed from that point forward. It might be helpful to think of plant growth in terms of a brick wall, with construction taking place brick by brick. A brick might be nitrogen, chemical symbol N, and we can compute the requirement. We can also compute the need for phosphorus and potassium and other blocks used to build a plant, atom by atom.

There are five million atoms in a drop of water, perhaps even more. At, say, 60 degrees Centigrade, all atoms should be the same size. This fact provides the insight that if you have the right drop of water on a crop, there is a terrific amount of energy in that package. It takes a pint of water to make one pound, and each pint has 10,000 drops. With this knowledge we can easily compute the number of drops in a gallon — 80,000 drops in the eight pints that make up a gallon. Once this habit of computation becomes a way of thinking, fertility management takes on a new dimension. The active ingredient to be used is often not even a teaspoonful per acre. This knowledge sets up another question — how many drops in a teaspoon? The answer is approximately fifteen. How many teaspoons to make a tablespoon? Approximately three! So the computation for drops in a tablespoon becomes a matter for simple arithmetic — 45 drops. So does the reckoning for atoms in that same number of drops — 15 x 5 million, or 75 million. Now we need to know something about the soil in the top six inches of those 43,560 square feet called an acre. Fortunately research has provided the answer — two million pounds per acre. A single drop on that acre would reserve only two and a half atoms per pound of soil. A teaspoonful of drops distributed over an acre bumps the number of atoms per pound of topsoil to thirty-seven and a half.

The principle involved here is simple in the extreme. Herbicide companies use it as a convenient tool to distribute a very small amount of lethal stuff over an acre of soil. Obviously, one drop in the middle of an acre is of no value whatsoever, regardless of the ingredient. In order to distribute a teaspoonful of material over an acre, water is required. Now the product becomes more effective because of how it is managed and handled.

In harnessing the Biological Theory of Ionization to crop production, it is imperative that you have this basic mental picture of dilutions to achieve uniformity of distribution even at one pound per acre. Unfortunately, most farmers are mentally conditioned to think that they need 500 pounds per acre because that's what they have been doing in the past. And, of course, if you put one pound in the center of an acre, that pound won't work either.

The secret is how to divide it up so that, say, one pound can cover an entire acre. So back to basics and the intelligence that there are 10,000 drops in a pound, and approximately 5,000,000 atoms in each drop. The equation is simple — 10,000 x 5,000,000, or 50 billion. These 50 billion atoms properly distributed over an acre suggest plenty of atoms for every square centimeter of territory in that 43,560 square foot area. In fact, distribution of a herbicide at that rate could well serve up a disaster.

The number of atoms per pound of soil seems astronomical, but it is also the key to understanding how to make fertilization more effective.

View this procedure, if you will, as you would an electric blanket. You do not want holes in the blanket because that would permit heat energy to escape. Our objective is to make an energy grid over the entire field, one that does not leak.

From sun to cell, with the atoms of creation as a common denominator, everything is related to everything else. We are told that the ancients knew this — in fact, Acres U.S.A. argues persuasively that the ancients along the Nile were great naturalists, and that they embodied many of their findings in that wonder of the world, the Great Pyramid of Cheops. To the north lies an area that has been farmed as long as Mainland China. For at least 6,000 years, nutrients have been bestowed on the delta by Nile floods carrying nature's own mix of carbon from Abyssinia via the Blue Nile and paramagnetic stone from arid Africa via the White Nile, mixed at Khartoum and stirred by the cataracts at Aswan — a top dressing rich in minerals and balanced to the magnetic flow from the equator to the poles.

This one case report gives us pause to wonder about the anomalies of planet earth.

In 1957 our first great spaceship venture came to fruition. That event kicked open the door to knowledge lost since the days of the ancients, if indeed they knew. At one point in that space flight Mission Control lost radio contact, and with the loss of contact they figured they'd lost astronaut John Glenn. Actually, Glenn was going through a radiation belt at the time.

In 1948 Dr. James Van Allen of the University of Iowa discovered this radiation belt, only to suffer ridicule. A few years later he was proved correct.

We live on a floating speck in the planetary system, on this earth being swung — if not by an umbilical cord, then on a gravitational string — in a 300 million mile orbit around its nebular sun. This planet wobbles slightly on its axis so that on what we now call June 21, summer arrives in the Northern Hemisphere, plants grow and food is produced.

The earth spins counter-clockwise in direction. This spin creates a magnetic field, or what we call gravitational pull. There is an anionic belt some 110 miles above the earth's surface. The exact distance is variable because the belt is elliptical in construction. This is simply a shield of negatively charged particles wrapped around spaceship earth like a security blanket. A spacecraft entering the earth's atmosphere through the Van Allen Belt at the wrong angle will disintegrate — which is why John Glenn lost radio contact, and why some of the shielding on the spacecraft flaked away. The farm connection to all this isn't even subtle.

At the stable end of that gravitational string, the sun for our local universe sends its bombarding beams of anionic energy into

ERGS

It takes only 25 pounds of feed to take care of 100 baby chicks for the first week, 75 pounds for the second week, and 175 pounds for the third. This same principle is true in crops. Young plants need very little food when the seed first comes up. As they grow older, they need more and more. So, you should put the plant food down in such a way that as the plant gets older, the availability of the plant food increases. This way you will not be under or over fertilizing, and thus not wasting money.

The term *ergs* designates a reading of how much plant food in terms of chemical energy is available per second per gram of soil. When planting, you should start out with around 40 ergs; this is the minimum you should have to even plant a seed. Then, as these plants grow, the ergs should increase as well. At the latest stage of growth, when production increases most rapidly, the desired level of ergs is between 100 and 200. However, it can reach as high as 400 for a few days at a time. With corn, for example, the time from tasseling and dying of the silk is the time that the ergs should be the highest in order to produce a maximum crop.

There are three basic substances that can be used to control ergs. The first is superphosphate, 0-20-0, which is not used for the phosphate but as an ionizer to release energy. Superphosphate is highly acid and reacts with the calcium in the soil to release energy. Second is ammonium sulfate, 21-0-0-24. It can be used to raise the ergs, as well as to add nitrogen in soil where the nitrogen levels have dropped. The third is ammonium nitrate, 33-0-0, which is used to raise the ergs in the spring in soils needing nitrogen. It cannot be used for an application in the later stages of a crop. Generally, 100 to 200 pounds of any of these products should be used, the specific amount being determined for each crop. Generally, 100 to 200 pounds of any of these products should be used, the specific amount being determined by soil tests.

Suppose the crop is young, and the ergs are low. You would only need to add about 50 pounds per acre of electrolyte, such as the materials mentioned above. That would make enough nutrient for the crop to do well. After that, another 150 pounds could be added, which would last throughout the maturing of the crop.

The ergs test tells what value is being gotten from the nutrient that is in the soil, but it doesn't reveal the source of the energy. It is very necessary to know why the ergs are there. For instance, ergs could come from sea water, but seeds would not even sprout in it. Just because a soil test reads 2,000 ergs at planting time doesn't mean the level is good or bad. Every seed that is put into the ground may rot. The soil analysis must be used to know how to make the ergs valuable to the crop.

The next problem comes when you check the ergs and they repeatedly register at 1,000, even in the spring. This may mean that there might be an extreme loss of energy that fast, so this energy would be going into the air. Some of this energy could be picked up by the bottom of a leaf, but it's almost like throwing money away if the level of the ergs is too high too soon.

If you record an erg reading of 1,000 and a pH of 2, this situation could be caused by the sulfur or aluminum in the soil. The aluminum in bauxite is what affects the ergs in this way. It is a very common condition in the state of Georgia. If sulfur is the problem, the soil will dry out. Aluminum will not do this. If you have this situation, we would suspect one of these two imbalances, because the pH is down. This is one time when it is important to know the pH. In this case, the way to drop the ergs is to add lime.

Another situation represents the other extreme, that is, when the pH is around 8, and you are recording high erg readings. In this particular case, a pH reading of 8 tells us there is plenty of lime, yet the soil analysis may read zero on calcium. This situation could be caused when magnesium is converted to a form that no longer enters into the reactions of the soil. If chlorine is causing the condition and lime is added, the chlorine will be released into the air by oxidation.

If the ergs in the soil are being created by elements that are not plant foods, they then are not counted in the erg calculations. For example, if you have very low, or 0-0-0 readings on your soil test, but show an erg level of 1,000, these are not plant food ergs.

50 — Too Low, Growth Stops
  — 1,000 — Beware.
  — 1,200 — Problems Start.
  — 1,500 — Major Problems.
  — 1,800
2,000 — No Growth.

earth. It is this light that keeps the earth going. These entering beams of light are concentrated anions, or negatively charged particles. They strike the Van Allen Belt as they enter and are deflected a bit, causing them to hit the earth's surface at an angle, putting it into a spin. This spin generates gravity, which has the function of pulling energy toward the earth. Out past the Van Allen Belt, there is weightlessness and a lack of gravity for almost all practical purposes. There is no time and there are no seasons. It is pitch dark out there except for reflections from the sun, moon and earth. This arrangement is our source of power for the purpose of crop production. Fertilizers increase and decrease the charge in specific atoms. A nitrogen atom or a phosphate atom has a number of anions and cations or electrical charges. This is the basic understanding we invoke when we work with the LaMotte instruments now common in mainline farming for Century 21. In short, our purpose as farmers is to measure anions and cations within the atoms of the fertilizers used to produce crops.

Under the aegis of the Biological Theory of Ionization, we will build a jigsaw puzzle in which each piece will have as much meaning as the mosaic of the whole. We will arrive at a new frontier. We will discover and compute the shape and tint of the missing atoms, and have a way of finding the answer for timely insertions into a fertility regimen.

The north magnetic field starts below the Hudson Bay in the Eastern Standard Time Zone. All of North America is in the north magnetic field. The actual magnetic fields do not flow from the North Pole to the South Pole. They flow from the equator in a circle as indicated in the diagram on the next page.

Those who want some further understanding of this phenomenon have only to read Genesis. If that wonderful and sometimes esoteric language is to be understood, we must note that the earth's axis and magnetic fields are now off center from what they were before the Biblical flood. It is this wobble that accounts for the summer and winter solstice, the seasons and the rhythm of much crop production. Carey Reams was a competent Biblical scholar. His studies and insight revealed to him that before the flood they didn't have seasons as we know them. Those who have

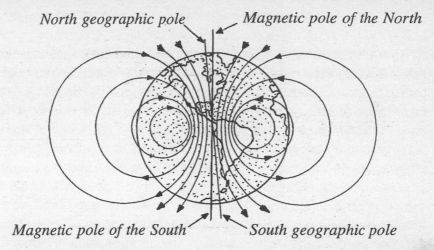

North geographic pole   Magnetic pole of the North

Magnetic pole of the South   South geographic pole

the interest and inclination to pursue this line of inquiry will want to read about Noah and the two or three Biblical chapters that expand on the subject.

Magnetic fields are very important from a cropping standpoint. Take the matter of row crop planting. Farmers have often noticed that corn or soybeans planted in east-west rows outperform the same crops planted in north-south rows. This has to do with the magnetic field. Normally if you plant in an east-west row, the roots will spread out toward the north. Root growth will proliferate faster, accounting for a higher yield. On certain plants, root systems are more extensive to the north in the North Temperate Zone, and the fine hair rootlets will be far greater to one side of the plant than to the other. Reverse the magnetic field in a given area with iron filings, taconite and a little compost, and the roots will have a tendency to grow the opposite way. In other words, you can grow roots in a direction opposite to the equator to North Pole magnetic stream by erecting a mini-magnetic field.

A mini-experiment will prove the point. Take a watermelon or pumpkin plant, and spread a fine line of compost for the roots to follow. It will be possible to run a root system 100 to 150 feet in one direction. Everything is related to everything else. The speed of light is approximately 186,000 miles per second. The speed of atoms traveling through the roots of plants is no snail's pace either. In fact we compute it the same as the speed of light. There is no reason to believe that a 100-foot trip at 186,000 miles a second will compromise the growth of a plant.

Again, the strongest magnetic pull is at the pole, meaning that there is more energy at that same pole. At the latitude of Mexico City, there is less energy. And for that reason 110 day Iowa corn takes days longer to mature at that south of the border location. Even without hands-on experience, reason will tell you that it will take longer to grow corn — same variety and number — than it takes to grow the crop in the cornbelt. In central Mexico it takes approximately nine months to bring a crop to maturity, whereas in the cornbelt it takes only 110 days to grow 110 day corn, proper fertilization being assumed. This growing season can be shortened another ten days — perhaps more — by creating a more powerful magnetic field on that acre of land. More atoms per acre will govern that energy pull if there are no holes in the blanket.

The markets are not entirely comfortable with the prospects discussed here. This business of cornbelt farmers shortening the growing time for certain row crops fouls up the candy factory. The trades have their set schedules, and their futures options have fixed terminal points. When the farmer installs his electronic blanket and turns up the charge, the crop may mature too soon for trade comfort. When this happens, the farmer should be prepared.

Thus it appears that in addition to death and taxes, the farmer can count on a natural magnetic flow and on adjustments he can make. In Alaska, unlike the latitude of Mexico City, corn will mature in 45 to 60 days, because as the magnetic lines converge near the North Pole, the crop grows faster. The shadow of the Pole by itself isn't enough to push 110 day corn to fruition in less than two months. Nutrients are required, and any shortfall can cancel out the optimistic expectations of corn at less than 110 days. But there is one shortfall that affects crop production worldwide, namely soil carbon — for it is the depth of the carbon in the soil that determines the depth of the magnetic field. It is this carbon connection that makes one 43,560 square foot acre different from another, and explains why a fertilizer will work on one acre and not on the next. Carbon requires our attention, our study, and its character, our measured response.

3

THE SYMBOL "C"

Paramagnetic stones and soil particles attract. Diamagnetic particles repel. The term of choice is flocculation, a property that causes soil particles to cling together. Flocculation is destroyed by long periods of drought and by a wasting away of the magnetism in the soil. The quality of this magnetism is governed by carbon. That little atom, symbol C, atomic weight 12.0111, has to be reckoned with if real estate is to be kept from changing hands on the wings of the wind.

Farmers and their fertilizer consultants frequently do not appreciate the fertility link with carbon and its role in setting up a magnetic field. But without it, topsoil will not stay at home.

There was a time before the age of killer chemicals when free real estate — bestowed by the inept practices of the neighbor — was something to be wished. Today, one farmer's folly is not necessarily another farmer's blessing. There are many chemicals of organic synthesis that should never be collected by air.

A few seasons ago my lawn was covered a quarter of an inch deep from a neighbor's field. This dust was laced with one of these herbicides. It annihilated the grass and for all practical purposes set up black advertisement to the effect that I didn't know what I was doing. Clients and visitors shook their heads in abject disbelief when they pulled into my driveway. It wasn't until late summer that I discerned what had happened. Toward the end of summer, I finally got the grass on the right track and moving ahead. The lesson here has to do with the role of carbon. Not only do we have to increase the carbon content of our own soils, we have also to encourage our neighbors to consider a new mainline agriculture for our own survival.

In many cases, farmers can do quite well without working on their carbon connection. But if they're going to hold that magnetic field and structure a fertility program that won't debilitate the bank account simply to get a crop, then the question must always be asked — *Where can I add to the carbon?* and *Where can I find products that contain carbon?*

This carbon requirement is not an illusion. Indeed, I can cite chapter and verse from now until next month illustrating the point. One should do. I have in mind a gentleman who studied the Biological Theory of Ionization fully a decade ago. Recently, he harvested between 200 and 219 bushels of corn off each of his 1,000 acres. But not until he learned how to handle carbon could he make the system fly.

When we increase carbon, the soil becomes more magnetic. The prime source of energy arrives from the sun. It is free. It confers energy ion by ion and figures in the business called ionization — the process of putting together and taking apart atoms of fertility compounds.

It is our function as farmers to *take apart* and *put together*. If we have a healthy soil, we need to take it apart so that we can reassemble certain atoms in the plant. We have to break out the raw organic matter. This is magnetic, of course, which means that it is trying to hold itself together. As a consequence, it becomes our assignment to create a magnet that is stronger than the magnetic force in organic matter.

Let's assume we have corn stover in the rootbed. This stalk has a magnetic charge. At one time this charge was superb because it accomplished the task of building a stalk. It was able to attract. Now the carbon in the soil has to pull it apart. That soil carbon has to be constructed by bacteria as amino acids. The sequence for action is at once simple and complicated in the extreme. Bacteria have a stronger magnetic force than the corn stover. As they break down the corn residue, they lose their electrical charge. In a weaker form the breakdown product becomes an amino acid first, finally carbon.

In order to make an amino acid, *carbon*, *nitrogen*, *oxygen* and *hydrogen* are required. These aminos are the workhorse labor force in any soil system.

Inserting bacteria into a soil that has no food for this work force is futile. Most of the new population will simply die. That is why any successful inoculation program has to deal with the business of building a home for bacteria and stocking the shelves. Microbial workers always obey the Biblical injunction to increase and multiply — if they can. Thus the soil life profile is always "in process." Some components in the soil system are always being taken apart, and some are always being put together.

The problem is that under cropping conditions, many soils become lopsided.

Possibly I can illustrate the requirement for this coordinated taking apart and putting together by citing a study made by I. G. Farben Company and the German WWII war machine. The Germans knew how to put that machine together, but they had problems with parts. Rubber came from the tropics. Ingredients for gunpowder came from Chile. Thousands of ocean miles separated raw material sources from the "putting together" factory.

The ionization process that governs crop production can be understood via analogy, abstraction and the arithmetic of the atoms. Taking the first, ionization is similar to the activity in an electroplating tank. You have a negative pole and a positive pole. Usually, powdered silver is introduced into the solution, with current taking on the role of guiding the plating material to the object to be plated. Usually a stirring device keeps the plating powder in suspension. Still, no plating will take place if the solution contains no electrolytes.

Carey Reams sometimes told the story of an enterprising young man who got into the electroplating business in Orlando, Florida. He had worked in a plating plant and through the years he had become tantalized by the profits that came to the business. So the young fellow purchased tanks and silver and otherwise invested in the operation. On job after job, he turned on the current, and nothing happened. The lenders became disenchanted. Not only didn't the business generate income, it failed even to service the debt.

Reams was hired to determine why the equipment wouldn't plate. Reams said his fee for troubleshooting the operation would be $10,000. To sweeten the deal, Reams offered to let an unbiased third party hold the fee, in this case the local sheriff. Reams said, *I will guarantee now that absolutely this will work — and if it doesn't work you don't owe me anything.* Reams then went to a grocery store and brought back a brown paper bag. He poured the contents into the plating tank. It was common table salt. For electroplating, you need an electrolyte. In a soil system you need a major electrolyte, namely nitrogen. There are others, but nitrogen is the key. It is a requirement in every plant cell. If nitrogen is in short supply, the plant cannot build new cells.

You may run into a situation in which a nitrogen form will not work. The only advice is to look at another form. The bottom line question has to be, *Have I met my nitrogen requirement for that crop?* And a sidebar for the question countdown should be, *Is the farming operation going to be limited to one application of nitrogen, say, a preplant in the spring for whatever crop?* If so, whatever the amount indicated, it must be increased by 30 to 50%. That is a certain way to stay out of trouble.

If you really want to have a better understanding of crops and plant growth, then read a good encyclopedia's entry on electroplating. Withal, soil science is not an exact science, even though we presume to invoke delicate calibration. A tremendous number of variables intervene to impose trial and error status on much of what we do. Every single acre of soil, just like every person in a room, is unique. When a university or a commercial laboratory sets up a test plot on a given acre of land, the result is good only for that particular acre, and it does not apply down the road. Other acres in other areas may respond quite differently.

Two terms Dr. Reams used were anions and cations. These terms are very familiar to someone who has studied chemistry. The problem is Dr. Reams attached a different meaning to terms used in chemistry. These terms are describing elements from an *electrical* point of view, not from a wet chemistry description. For example, calcium is a *cation* in wet chemistry, but it is an *anion* from an electrical point of view. Therefore, when you see these terms, do not think of them as you would in wet chemistry.

An anion spins in a clockwise direction. It comes from the sun. It passes through almost any object. It passes through water as well as a brick wall. The anions given off by a burning log in a fireplace can pass through the human being warmed by the heat. Sound can go through a wall that isn't insulated enough for buffering effect. In other words, a group of anions in a set pattern becomes an audible voice to the receiving equipment, the ears. Public speakers and schoolmen lose a tremendous amount of energy when speaking before an audience, far more than if they were doing physical labor. It has been computed that an hour of lecture time is the equivalent — in loss of energy — to eight hours of hard physical labor.

This anion, then, has an assigned energy value for purposes of communicating with other ions. Reams defined these values as Milhouse Units, and he used these units to compute the arithmetic having to do with atoms. We're not certain whether this is a coined term that never quite made it into scientific dictionaries, or whether its author has been merely ignored, as has the Biological Theory of Ionization itself. We do know that this arithmetic works, and this supports its reason for being.

According to this scale, anion values range from one to 499. Anions always spin in a clockwise direction. This spin and the Milhouse Unit values proposed by Reams dovetail to provide a method for computation of energy in fertilizers with a common denominator suitable for field application.

A single anion is much smaller than an atom. It is smaller than an electron. It is what makes up an electron. There are fertilizers that have a tendency to be anionic. Applied to a crop, they cause prolific growth of foliage, but they do not produce the part of the plant the farmer usually sells, the seeds and the fruit. It is possible to manage a corn crop for silage so that no seeds set whatsoever. This can be done by putting on the right anionic fertilizer at the right time. This is not the usual intent, but it can be the end result. Much the same can be the unfortunate result with beans, namely prolific vines and few beans.

This intelligence has solved one of the military's greatest nightmares — communicating with submarines below water. Based on the propensity of anions to penetrate walls and water, scientists have developed anionic energy wavelengths that can communicate with submarine receivers in the depths of the ocean. Parenthetically it must be noted that if a surveillance team wants to keep track of someone, a receiver somehow attached to this person would make hiding impossible.

A cation is the next smallest unit of measurable energy. It has an energy value on the Milhouse Unit scale of 500 to 999. It always carries a positive charge and always spins in a counterclockwise direction. There are some fertilizers that are primarily cationic. And there are some crops best influenced by cationic fertilizers — cabbage, for instance, for which seed production is not sought generally. A highly cationic fertilizer will cause the crop to bolt and go to seed — and with that the crop is gone. Cations won't pass through that proverbial brick wall, but they do pass through the water molecule.

He who controls the anions and cations controls the universe. Insight and knowledge have made a measure of manipulation possible, but control for all practical purposes is out of the question.

A summary of sorts is now in order. First of all, anions and cations spin in opposite directions. Anions can switch to becoming cations if they reach energy values over 500 Milhouse Units. At this point the anion flips and changes the direction of its spin. This happens during the growing season, usually in August in the cornbelt. If it did not, there would be no seed crops, and in most cases the farmer would have nothing to sell.

Most soils are switched during the winter months to anionic, and during the summer to cationic to set seed. Unfortunately, there are seasons during which there is not quite enough switching under the management plan in effect. It is often too cool. A strong cationic spray is indicated. But because of a lack of knowledge, repair measures are not taken and the crop yield suffers.

There are many dimensions in our discussion of energy. One and all, they will be unfolded and examined in the chapters that follow. By definition, carbon will remain central to our discussion, even while we measure anion and cation fertilizers, and their impact on each other.

We must never lose sight of the fact that life is really organic. It is built on carbon. Managed unwisely, salt fertilizers can destroy the carbon in the soil. Such destruction carries with it the possibility of hormone and vitamin destruction in the reproductive parts of plants. Thus the capital of life is also destroyed from one generation to the next if carbon in the soil is annihilated by unwise use of chemical salts.

The conductivity meter measures total ion concentration times mobility value, and therefore provides a fast response digital readout for soluble salts. It is also an ideal instrument for evaluating water quality, plant food availability during the several phases of growth, and viability of seed germination. This instrument is a valid tool for the move into mainline farming for Century 21.

4

THE KEY TO BIOLOGICAL LIFE

The aim of science is to foresee. Science describes phenomena and tries to join them by laws, and these laws enable man to predict.

Through the study of the motions of the heavenly bodies, for instance, astronomy has succeeded in establishing laws which enable man to calculate the position of these bodies with respect to each other in the future. In like manner, physics and chemistry describe the behavior of solid, liquid and gaseous bodies, and these descriptions lead to laws which replace the amazement of ignorance by the sureness of knowledge. When it has been experimentally observed that certain phenomena seem to be invariably linked to the first by a relation of cause to effect, this observation is worded in such a way that it enables a man to predict these phenomena quantitatively or qualitatively whenever the same conditions are present.

In philosophy courses that are rarely taught nowadays, Aristotle is cited for his classification of science as *practical* and *speculative*. The first aims primarily at complying with the aim of science mentioned above — to foresee. Speculative science seeks the discovery and understanding of the order in the universe, the order of nature as it exists independently of man. Aristotle divided this brand of science into physics, mathematics and metaphysics according to the degree of abstraction each discipline supposes. Many people have a problem with abstractions. As a consequence, they are forever thinking and talking at cross-purposes. When we talk about anions and cations, we have to invoke the appropriate abstractions. After all, we can't see an anion. We are obliged to create a mental picture of clockwise and counterclockwise spins for cations and anions. It is clear, then, that what is called scientific truth can be accepted only in a limited sense, and cannot be considered literally in a system of absolutes. Scientific laws are all *a posteriori*, forever governed by the facts to which they must submit.

In the process of sorting out answers, we have to remember that we are neither oracles nor God, even though the device called abstraction hints and commands at understanding well beyond anything we can observe. When a yield is missing, it is the abstract view that tells us the soil is more negative than positive, and the amount of plant food is more anionic than cationic. This habit of handling abstractions will enable a farmer to view seed corn as a container full of anions and cations in its simplest form, spinning in such a pattern that we perceive it to look like corn that we have identified and tagged with a name. Everything we see, when there is a concentration and a specific configuration, is composed of anions and cations. An anion is the smallest amount of measurable energy. A cation is the smallest amount of measurable energy in positively charged elements. In the right configuration we wind up with an atom of an element — copper, zinc, cobalt, potassium, nitrogen, whatever.

Every living plant owes its biological life to the *loss of anions and cations*. If there is no loss or gain going on all the time, there is no life. The plant must gain anions and cations while it is growing. Later, after crop residue has been incorporated in the soil, decomposition requires a similar loss. In short, plants live off the loss of anions and cations in the soil from the breakdown of crop residues or the addition of fertilizers that — as they give up their energy in the form of cations and anions — allow the plant to grow.

If a soil becomes perfectly synchronized, no energy exchange is possible. At this point the crop stops growing. Fortunately there is a higher power that does not permit this to happen.

At ground level, and from our point of view, the chief impediment to such an energy synchronization is temperature. During any 24 hour period anywhere in the world, there is a variation in temperature of at least a few degrees. If you have a solution, the addition or subtraction of heat will force molecules to speed up or slow down. In August, when it is extremely dry, molecules can move too fast and fly right on by the rootlets. If there isn't enough magnetism to pick up these nutrient molecules, then there will be a shortfall of the anions and cations required for plant growth.

It is a little like boiling water. You can add so much heat to the water that the molecules literally explode. The water evaporates. This is a serious problem in some soils. Carbon governs the release of energy in any soil. If the carbon content is low, there is a greater need to fertilize. A low carbon content according to field instrumentation usually means a lower plant food recommendation per application, but it also means more applications. Too much fertilizer means too much heat — too much gasoline on the bonfire, so to speak, all "blow" and no "go." Forty-eight hours later the fertility load is lost into the air.

This means that in low carbon soils, you have to put on less more often, or you have to use slow release materials so they will carry longer in the low organic matter soil.

One of my favorite formulations for small grain crops is a mixture of 28% nitrogen Thio-Sul 12-0-0-26 and molasses. This works well for about two years, at which time calcium must be checked. I had one farmer try it on corn with impressive results. This is called putting nitrogen on in amino acid form. When using this mixture, nitrogen stabilizers that kill soil microbes are not needed. It is nitrogen that puts the "sticky" in glue. If there's no nitrogen it won't be sticky. Think of it this way when making up a formulation. Did you ever drop your chewing gum in the sand when you were a kid? Which was easier to remove — the sand or the chewing gum? A mixture containing nitrogen is essentially like dropping chewing gum in the sand. Moreover, it propagates itself once it is out there with the sun shining on it. It expands many times. And, as applied in the field, it has brought in unbelievable yields, chiefly in small grains, although I have had cornfields that made one sit up and take notice.

In addition to temperature, moisture will keep a soil from total synchronization. Moisture will fluctuate considerably in any topsoil. It cycles up and down in the soil over a 24 hour period. Proof for this statement can be assembled at the cost of a few hours of sleep.

Go into a field at 3 o'clock in the morning and check it out. Then return to the field at 3:00 p.m. when it hasn't rained. One thing you will notice is that no rain does not necessarily mean it hasn't rained.

A walk to the field at 3:00 a.m. — especially in June — can result in a return to the house soaking wet. The more carbon in the soil, the wetter the return trips. A few hours later, it would be possible to walk through that same field and never discern the visit of that rainless rain. Many farmers have learned how to fix a crop in the absence of rain. Carbon in the top six inches of soil controls water's capillary return.

This use of carbon to keep fertilizers in the top six inches of soil has become the key to making many midwest farms successful in growing crops. Once this principle is established, then it is no longer necessary to distribute tons of materials on every acre of topsoil.

Carbon holds approximately four times its weight in water. In other words, one pound of biologically active carbon can absorb four pounds of water. The key words here are *biologically active carbon*. Only this type of carbon has the capacity for absorbing moisture. A one pound diamond — which is pure carbon — will hold no water whatsoever. By way of contrast, a natural sponge

ABSTRACTIONS

The process of abstraction admits various degrees according to the depth with which the mind penetrates into the data of experience. Aristotle divided speculative science, physics, mathematics and metaphysics, according to the different degree of abstraction which each supposes. The degrees of abstractions are defined as follows:

First degree of abstraction. This is the abstraction required by the physical sciences. It abstracts from the determined individual matter of opaque objects, of sensible change, until the intellect confronts the universal essence. From the knowledge of these essences the mind is able to grasp the laws which govern the physical, material world and are universally applicable to the objects of sensible experience.

Second degree of abstraction. This degree of abstraction goes beyond sensible changes and discovers a continuing element in all corporeal beings. The intellect abstracts not only all individual and sensible qualities, but even from all sensible matter, as is seen in the concept of a geometric triangle or a geometric circle. Only a form with its relation to the intelligible is perceived. Whether the circle, for instance, is gold or silver is no longer considered. Aristotle saw this abstract notion of quantity as the proper foundation for the science of mathematics.

Third degree of abstraction. This degree of abstraction reaches not only beyond sensible qualities of change and of individual matter, not only beyond quantity and sensible matter as well, but even beyond intelligible matter to what Aristotle calls ultimate reality. The ultimate reality common to all is *being*.

Actually, the term *metaphysics* was not used by Aristotle. It was coined by Andronicus of Rhodes (70 B.C.) who collected the works of Aristotle and placed this portion of the writings of Aristotle after the books of physics, hence the name *metaphysics*, meaning "after the physics." Aristotle himself called this portion of his writings *theology* because it reaches to the study of God, and *first philosophy* because of its necessary importance in explaining the First Cause. Metaphysics was in turn mutilated by Christian von Wolff into distinct sciences designated by the curious titles of *ontology, cosmology* and *theodicy*. The ability to handle abstract thought is what made a man like Carey Reams take cause and effect up the ladder to his Christian beliefs and to God.

— which is natural carbon — will hold even more water than four to one. At any point in time, there is a temperature change. This improves the moisture holding capacity of the carbons, and forecloses the possibility of soil synchronization — that theoretical position of balance when anions and cations no longer exchange and crops cease to grow.

During the growing season it is not uncommon to get two or three hot days in, say, August. If the soil is low in biologically active carbon, the corn will start to curl and wilt. This means the plant is in negative energy balance. It is deionizing instead of ionizing.

Even in the heart of a drought, there will be corn plots that do quite well. Active biological carbon in the soil makes the difference. Just as the ocean has a tide, so does the cornfield. The gravitational pull of the moon causes moisture to rise in the subsoil. So does atmospheric pressure, which allows the soil to take a deep breath, once a day, exhaling and inhaling much like a human being, putting moisture on the ridgehill and monitoring plant uptake. If a carbon product will hold water, the farmer is entitled to great expectation. Soft coals and new coals called humates are highly absorbable and possessed of the correct spin, but a little goes a long way. The need for humic acids in a soil is very small. Too much is worse than no application at all. When humic acids are applied in liquid form on, say, the worst sands in Arizona or New Mexico, a gallon would be too much. Most of the time a pint to a quart per acre would be indicated.

Carbon application depends on the specific type. Molasses calls for one rate, sugar another. Water movement has its nuances. Hillsides tend to have less carbon due to conventional management practices, and on bottomland soils it is often worse because more water will be pulled there because the greater amount of carbon will result in greater electromagnetism.

This much stated, it should be at once apparent that there is a great difference in tilling in the morning and afternoon and late at night. The phase of the moon determines whether the soil will be loose or tight. Also, applying a fertilizer based on the phase of the moon will prove far more effective when the soil is low in organic matter. When the soil is high in organic matter, the moon's phase will make no difference. By applying a fertilizer when the soil is looser, it will be far more effective than when it is more compacted, again based on the moon. When there is a full moon, the soil is more porous, even if it is low in organic matter and has a tendency to compact.

The moon is a dear teacher, if we have the wit to learn. I have a young son, a typical American boy. One day he got hold of our lawn fertilizer spreader. I had a half bag of calcium nitrate sitting in my garage. The boy filled the spreader and went to work on the lawn, obviously not in a straight line. He made a circle one way, and then traced out erratic paths until the hopper was empty. Where I sit to have my breakfast in the morning I look out that window. Within two or three weeks, green strips appeared all over the lawn. They were there all summer. I confess I chuckled every morning thinking of my son making that path. It also provided me with a learning experience. I thought I could fill in around that pattern with fertilizer. I spread out the fertilizer and I didn't change the spread setting. I may not have walked the same speed as my son, but, where I spread in the fertilizer, the grass burned.

Finally, I got on the phone with Reams. He said, *You mean to tell me that after all the training I've given you that you don't know what's wrong?* I said, *No*, and he said, *Well, I'm not going to tell you either. You're going to figure it out for yourself so that you will remember it the next time.* So I sat around and looked at that lawn for another month, and I really couldn't figure it out. I called him up and he gave me a hint. He said, *You go back on the calendar and check the phase of the moon when the boy applied the fertilizer. The soil was lighter and fluffier at that time.* My lawn is ninety feet of pure clay straight down. There is no topsoil. He convinced me that there is a difference according to when you apply calcium nitrate. I learned a valuable lesson from that. It is always well to remember, when you're comparing notes on what somebody else did, and you go home and do the same thing and it doesn't work, make sure you've taken all factors into consideration. By the time you find out what you need to know, the moon may have changed, and the land's hidden hunger may have been modified.

You can see hunger in a soil and you can feel its force when mud sticks to your boots. Such a soil seems to grab anything. It almost shouts from the rootbed that it needs to be fed. Our abstract point of view tells us that the cation exchange sites are empty. Carbon and calcium are needed. Only serious imbalance will cause soil particles to magnetize to your boots.

Some few years ago I encountered these phenomena in Indiana. The field in question was planted in corn. You could walk the length of those plotted acres barefooted, and your feet would remain clean. It was a lot like walking through a forest for refreshment. The soil did not take away energy, it conferred it on its pedestrian visitors.

Many of the lessons in this book will be strong meat, and the taste will be bitter to those who are afraid to be different. There is a saying that is often repeated in *Acres U.S.A.* — that farmers must think for themselves without reference to a higher approved authority. The authorities that seem to be approved nowadays are totally secular, often pseudo-scientific, and their rejection of new ideas usually starts with "There is no scientific evidence. . ."

Those of us who remind ourselves that a higher power is in charge recall Biblical allegories, such as the report on Noah. He was "different." He labored for 120 years building an ark and telling his world that it was going to rain. Then and now, his would be considered a ludicrous position. There wasn't even a warm-up rain until the legendary flood.

We mention again the Milhouse Unit, a system for graduation of anions and cations, and we will — in these pages — compute the units of energy in a single molecule of water. In theory each element could be scaled with Milhouse Units according to properties recorded on the Mendeleyeff table of elements. In fact, we will deal only with averages, the average value for any anion being 250, and the average for a cation being 750. The capacity for carrying current is what differentiates an anion from a cation.

If you go to that periodic chart of elements for a reading on hydrogen, the entry will reveal an atomic weight of 1.

This means hydrogen has one electron spinning around one nucleus. The spin can be either clockwise or counterclockwise, and hydrogen can be either positive or negative. Our Milhouse scale of values tells us that counterclockwise motion means a positive charge, a cationic energy with a value of 500 to 999. Hydrogen has a singular nucleus. If the outside is positively charged, then the inside is negatively charged, or vice versa, in which case the value can be from one to 499 for a single hydrogen atom. In our

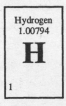

Hydrogen
1.00794

H

1

computations, we have a maximum for the anion. The total maximum energy in one cationic hydrogen atom is 1,498. The minimum is 501.

Fertilizers are not precision tools, graduated as it were to 1/100th of an inch. They may be processed with slight differences, and these can be significant enough to affect the yield of a crop. Simply stated, there may be less or more of the cations and anions in the electron shells of the specific elements in the fertilizer.

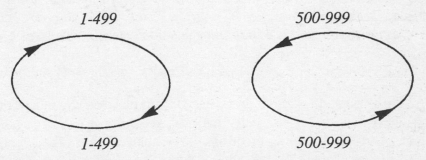

The good news is that we have a way of measuring these energies. And still there remains the riddle and mystery of soil science — how to determine whether an element is anionic or cationic when experience tells us that property can be either. If hydrogen rated attention as the only flip-flop element, there would be no problem. Although we have raised the question, we will defer the answer until more groundwork has been put into place and leveled.

For now it is enough to point out that the energy created by the fertilizer makes the difference. If a fertilizer cannot give up its energy, it is worthless for crop productions.

The pH meter is a valuable tool in eco-agriculture, but it does not have in its scale a reading for pounds of energy in a soil system. It will hint at whether the soil has anionic or cationic elements, but cannot furnish information as to which one is in control. The pH of a soil might read 7.4, suggesting its anionic character, but a trip to the same field in the middle of the night could easily turn up a reading of less than 7.

The point here is that each soil is unique and has definite requirements. It is the function of our brand of fine tuning to zero in on specifics for maximum crop production in terms of quality and quantity. Any molecule or atom will seek synchronization halfway between the lowest and the highest value for anions and cations on the Milhouse Unit scale. Halfway between 500 and 999 on the cationic scale is 750. Halfway between one and 499 on the anionic is 250. Thus, an overall average value would be 1,000, this for one single atom of hydrogen. Using this concept and these numbers it is well within the reach and capability of the farmer to calculate a number of different fertilizer materials on an average basis.

A water molecule is H_2O. Oxygen has an atomic weight of 16.

$$H - O - H$$

This means an oxygen atom has sixteen electron shells and one nucleus. Oxygen is normally cationic, meaning it carries a positive charge in the outer electrons. If we take 16 times the average of the cation, or 750, we have a formula — 750 x 16 = 12,250. That's the amount of energy in an electron shell of one oxygen atom. We also have one anion at the center of that water molecule. If we take an average value of that anion, or 250, and add it to 12,250, the sum becomes 12,500. That's the energy value for one oxygen atom. This water molecule has two atoms of hydrogen and one atom of oxygen, and is written H_2O. We have already computed the average hydrogen value to be 1,000. Since there are two of them in the H_2O formula, multiplication by two is indicated, and the total is 2,000. With the addition of 2,000 to the computed value of oxygen, the new sum is 14,250. That's the average energy value for a single molecule of pure water that contains two hydrogen atoms and one oxygen atom.

Just as rain is not necessarily rain, water is not necessarily water. There is dry water, heavy water, wet water. Those who travel a great deal have encountered motel water so dry that towel drying after a shower is hardly necessary. In another area the water may be so wet it takes two towels to dry off. The driest water may mean 501 twice in a computation, or 1,002, and the lowest value for a single atom of oxygen would be 500 x 16, or 8,000. Two hydrogen atoms at a 501 value — or 1,002 — added to 8,000 equals 9,002. The difference is at once apparent when we compare 9,002 to the previously computed value of 14,250.

If water is aerated as it plunges over a cataract or waterfall, it has a different effect than 9,002 unit water computed according to theory. Before the High Aswan Dam was constructed, water for the Nile and its Delta floods was alive. Together with carbon and paramagnetic stone, this magnetic and nutritional fix allowed Egypt to feed itself. Now, two decades later, Egypt has to import fully 50% of her food.

Different fertilizers and different molecules of water have different energies to release. That's why the Biological Theory of Ionization asks us to rely on instruments, and why one product has a tendency to work better in a certain situation than another, even though they both have the identical analysis.

One case report should illustrate the point. In Idaho, under one pivot of wheat, the yield is always 40 to 50 bushels more than neighboring pivots. Laboratory tests have found nothing different about the irrigation water used. One Biological Theory of Ionization student measured the water with a conductivity meter and made comparisons with water from nearby pivots. It turned out that the well servicing the high yielding pivot had a higher conductivity reading than other wells in the vicinity. So here was a factor the conventional laboratories could not explain with any scientific rationale. And yet the system discussed here had the answers.

Vapor pressure affects the energy in water. So does a radioactive charge. In the case report mentioned above, there was indeed a higher radioactive charge than in water from any of the other wells serving pivots. One way crop growth can be enhanced is to spread uranium, taconite, even iron filings over an acre of land. Granite dust also works well. All these things influence the charge.

To set up these several questions is to suggest that there is an answer. It is the aim of science to test for those answers, and then to foresee. A hundred years ago, scientists were few, and the scientific spirit was alive and well. Today, much of science is merely procedural and legalistic, and the spirit of pure inquiry has passed back to the font of its origin — to the farmer who accounted for observations, wisdom and agriculture in the first place.

5

CATIONS AND ANIONS

$E = mc^2$.

And behind that equation, Einstein's Theory of Relativity burst upon the scientific scene. When Einstein first came to Princeton, it was common street talk that only a half dozen masterminds in the entire country could understand relativity. All sorts of parables were used to explain it, including valid references to astronomy. But by the end of World War II, relativity was presented as classroom fare to sophomores in college. More than any development in science, relativity declared the cleavage between the evolution of empirical science and the lofty abstractions Plato set down in exact terms. Einstein, in his own book called *Relativity, The Special and the General Theory* [Robert W. Lawson translation] noted that "the development of a science bears some resemblance to the compilation of a classified catalog. It is, as it were, a purely empirical enterprise." He went on to point out that the actual process was much more complicated. This procedural business called science slurs over the role played by intuition and deductive thought in the development of exact science, he in effect said. The late William A. Albrecht said, "First, you have to have a vision."

Relativity basically sought to explain the swing of the stars and the mass of the universe, and anyone who seeks to understand the wonder and majesty of Einstein's insight soon meets up with minds on par with his own — from Mock, Einstein's teacher, to Eddington, Cottingham, Crommelin and Davidson — the people who went to Sabral, Brazil and the island of Principe (off West Africa) in order to monitor the solar eclipse of May 29, 1919, with Lorentz, Minkowski and Gauss added for good measure. Near total comprehension of the universe made near total comprehension of the mini-universe of the atom possible.

When Einstein worked out his equations he declared it possible to take matter and break it into electrical and heat energy. Rapid release of this energy provided first the theory, then the reality of the atomic bomb.

$$E_1(Heat) + E_2(Light) = mc^2$$

Translating the secrets of the universe into formulas suitable for farm application became workaday activity when Einstein, LaMotte and Reams discussed the problem only a few years later. Those five million hydrogen atoms in a drop of water could provide a good sized explosion if released all at once. That is what Fermi and Project Manhattan succeeded in doing over the desert of New Mexico first, then over the crowded cities of Hiroshima and Nagasaki. If it were possible to extract all the energy in a thimbleful of diesel fuel in a semi-trailer truck, theoretically the vehicle should be able to drive from Fairmont, Minnesota to Atlanta, Georgia. Obviously, we have not been able to harness all that energy. Too much of it goes out the exhaust pipe. Relying on his insight and understanding of mass and relativity in so-called space, Einstein learned how to take matter apart, secure with the knowledge that someone one day would be able to put it back together again.

A whole glossary of terms and concepts figures in life and growth. There is the magnetic field, source the sun. Carbon determines the depth of the magnetic field across an acre of land. The next item of importance to maintenance of a basic electrical field is manganese. If manganese is not present, seeds will not sprout. On the periodic table of elements, manganese is represented by the symbol

Manganese
54.9380

Mn

25

It has an atomic weight of 54.9380, which is one of the heaviest elements essential for crop production. With 56 electrons at an average value of 750, the bottom line is approximately 54,750. Nitrogen with an atomic weight of 14 has a value of 3,500, which is simply 14 x 250, the anion average. It takes at least 12 atoms of nitrogen to capture one atom of manganese.

Nitrate nitrogen is one of the main negatively charged elements in the soil, the other chief ones being calcium, atomic weight 40, and potassium, atomic weight 39.102. In terms of calcium, the result of 40 x 250 is 10,000. If each calcium atom gave up all 10,000 units of its energy, only four atoms would be required to capture one atom of manganese. No element will give up that much of itself, therefore a great many more than four will be required to capture that single atom of manganese. It may take as many as 15 or 20. If we have one pound of manganese per acre, it may take 500 pounds of calcium to serve up the energy needed to capture that manganese. A low test weight on a crop means that the soil was not working correctly to capture the necessary manganese. Certain symptoms can assist the farmer in reading the situation. If the soil is sticky, it may be so hungry the calcium can give up very little of its energy to capture the manganese. Soil that isn't trying to grab the boots off your feet will give up calcium more readily.

There is always a connection between soil, food and resultant disease conditions in the animal and human population. One of the reasons the U.S. population has one of the highest incidences of prostate cancer in men and breast cancer in women is the lack of manganese in the food supply. This lack suggests a shortage of all other elements because it takes the energy of nearly all nutrients combined to take on manganese.

Sources of heat energy are the sun and chemical resistance of the soil. To effect energy release in the soil, one anionic and one cationic plant food are required. The two give off the most energy. Too much plant food without suitable carbon buffering capacity can be disastrous. In other words, the lower the carbon content in the soil, the more critical the amount used for a single application.

Another form of energy is encased in the seed itself. Inserted in the soil, it starts to swell. If the soil can't expand with the seed, a problem for the growing season is in the making.

The next consideration is the electrolyte in the soil. The electrolyte is always a conductor of electricity — usually iron, copper, zinc, etc. The most important one is nitrogen because no crop will grow without it. Even if a cell needs iron, copper or zinc, it can't affect formation of the cell until nitrogen is present.

Management of a soil system is both an art and a science. If you add calcium to some soils, it will cause them to flocculate

BASIC PRIMORDIAL CELL

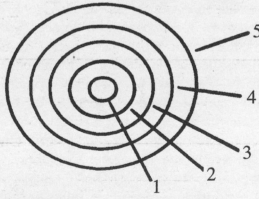

1. Nitrogen   4. Hydrogen
2. Carbon   5. Calcium
3. Oxygen   Phosphate is the catalyst

and fluff up, and others it will cause to set up like fast-drying cement unless carbon is added quickly. Sometimes the most valid lessons won't be found in agronomy manuals. I have an oil engineering book that shows how clay molecules stack together, and how they can be forced apart by adding certain elements. You can flocculate the soil, spread it out and fluff it up, or you can cause it to contract and compact. Oil engineers have developed useful information by making drilling muds. The right consistency for mud is all important as an opener. After that, the use of calcium and humates to increase and decrease the thickness of the oil drilling mud achieves an art form in the drilling fields. The same principles are operative in soil used for production crops. Calcium will flocculate some soils and cause others to harden. That is why a farmer really needs to know his fertilizer materials.

To get an energy release, it is generally necessary to work an anionic fertilizer against a cationic fertilizer. An example should illustrate the point. If you have one fertilizer that reads 250 on the scale, and one that is 750, and you run them together — one spinning clockwise, one spinning counterclockwise — moisture and carbon will control the rate of release. At approximately the 45th day of the growing season, when seed production is in full swing, a maximum energy release is indicated.

Earlier in the season, when branch and leaf growth are the objectives, it might be necessary to work with two anionic fertilizers. These should be different brands from different fabricators, since it would be highly unlikely that the two would have an identical amount of energy.

The same would be true when two cationic fertilizers are combined for powerful energy release. Different brands from different fabricators are most likely to have different energy loads on the Milhouse Scale, even though computation via averages would be invoked.

These are the three different methods of creating fertilizer energy — anionic and anionic, cationic and cationic, and anionic and cationic, all combined under the rules of the miniature universe that apes the great universe described by Einstein's relativity theory. Mixing these fertilizers is no plaything for the amateur. Especially when using dry materials, it is important to have a clear understanding of which ones are anionic and which are cationic.

FERTILIZERS

Ammonium Nitrate. 33.3-0-0 (anionic and cationic) NH_4NO_3. Ammonium nitrate is manufactured by reacting nitric acid with anhydrous ammonia, and is prilled and coated to prevent caking.

Ammonium Sulfate. 20.5-0-0 (cationic) $(NH_4)_2SO_4$. Ammonium sulfate is manufactured by reacting anhydrous ammonia with sulfuric acid.

Anhydrous Ammonia. 82% N (cationic) $2NH_3$. Anhydrous ammonia is manufactured by reacting hydrogen from natural gas with nitrogen from the air.

Calcium Nitrate. 15.5-0-0 (anionic) $Ca(NO_3)_2$. Calcium nitrate is produced by reacting nitric acid with limestone.

Nitrate Of Soda. 16-0-0 (anionic) $Na NO_3$. Nitrate of soda is no longer manufactured in this country, and small quantities of it in the natural form are being imported from Chile.

Urea. 46-0-0 (cationic) $(NH_2)_2CO$. Urea is manufatured from ammonia and carbon dioxide.

Ammonium Phosphates. (cationic). Monoammonium Phosphate 11-52-0 or 10-50-0; Diammonium Phosphate 16-48-0 and 18-46-0; Liquids 8-24-0, 9-30-0, and 10-34-0; Ammonium phosphates are produced by reacting ammonia with phosphoric acid.

Potassium Nitrate. 13-44-0 (anionic) KNO_3. Potassium nitrate is manufactured with potassium chloride, with sodium nitrate or nitric acid.

Chilean Nitrate of Potash. 15-0-14. Chilean nitrate of potash is a naturally mined product, imported from Chile.

Natural Organics (cationic); bonemeal 15-16% N; compost 2% N; cottonseed meal 6-7% N; peanut hulls 3.5% N; poultry manure 1.5-10% N; tobacco stems 2.5%-3.7% N.

Suffice it to note that a mix of a negatively charged fertilizer and a positively charged fertilizer requires immediate spreading. Delay will see the mixture first becoming too sticky for transport through the spreader, then set up like concrete. Still, an anionic-cationic mix is the key to growth and seed set. It is possible, of course, to make an end run around this problem by putting the dry materials in a slurry for immediate spreading. The energy and gas release in the slurry tank is beyond instant comprehension.

The value of the anion-cation computations lies in being able to fit each cog into the energy wheel in order to get the maximum response for the dollars spent. The Reams system takes the guesswork out of fertility management. Since the anatomy of weed and insect control is seated in management of the anion and cation exchanges, the ability to calculate energy values is tantamount to lessening and even eliminating chemicals of organic synthesis, the herbicides, pesticides, fungicides. This is no absolute statement. What works this year taking climate, weather, moisture and temperature into account might not work next year. But with the anion-cation energy scale, we have a system for indexing a specific season and plugging in a fertility mix that will work. In order to get maximum seed production, the plant has to be magnetic enough to draw manganese into it at the expense of other atoms in the plant giving up anions and cations. The DNA dictates the growth pattern of a plant. The question is always whether there is enough energy for maximum production. If corn kernels are denting, not enough energy has been pulled in. If there is enough energy, corn kernels won't dent, not even in hybrid corn. If the manganese uptake is sufficient for soybeans, flowers won't drop off. The plant will continue to set pods. After that there will have to be sufficient cationic fertility to deliver the beans set up by ample flowering.

It is rarely necessary to apply manganese. Most of the time, when you really get down to the bottom line, energy from calcium is lacking. An excess in the soil and a deficiency in the plant is a more believable scenario. In this case the real problem is not enough energy to take up manganese, in which case spraying on foliar manganese would not solve the problem. Calcium nitrate sprayed on very gently might help. At a four pound rate per acre in 20 gallons of water, such an application would be effective before or after emergence. Calcium nitrate would give a quick response, albeit only for a short time. Post emergence application in excess of four pounds per acre would likely result in a crop being fried, as in a skillet.

On a wet lab chemistry chart, calcium and magnesium are positively charged, but in electronic chemistry the opposite is true. This has been a problem to those who are not familiar with the grammar of the subject. When conventional soil audits evaluate the base exchange capacity, they always list calcium as a cation, and when college agronomists pause to look at the Reams Biological Theory of Ionization, they literally pounce on this apparent contradiction. Reams always annihilated their position by asking one question — *Does calcium carry electrical current?* Since calcium does not carry current, it has to be a negative.

Calcium chloride will pass a current because the compound has an electrolyte built in. One of the several rules Carey Reams codified as an integral part of the Biological Theory of Ionization was the one that says plants live off the loss of energy from nutrients during the synchronization of these elements in the soil.

The best way to illustrate that rule is to put baking soda in grapefruit juice. It will fizz like some miniature Vesuvius. This activity will go on until synchronization is achieved. To unsynchronize the process, it would be necessary to add more grapefruit juice or heat. The same principle is operative in the soil. The key to crop growth is prevention of total synchronization, and this is best regulated via the agency of carbon.

There are millions of organic type acids in the soil. One of the reasons liquid humates have not swept the field is that no fabricator seems able to make up the same solution twice in a row. The material works like an oscillating catalyst, always rotating one way or another, according to whatever forces are in the solution at the time of mixing, such as the temperature of the mix, and the ingredients used in the solution, water included. Thus the end product seems always to be consistently inconsistent. Technology permits manufacture and tagging of N, P and K in a consistent manner. Such standardization is seldom possible when most natural products are involved. Regulatory officials — who like to think of agriculture as an industrial procedure, and not a biological procedure — have used nature's diversity to hammer down many valid products and keep them out of the market. As it stands today, innovations of special value have to be buried in N, P and K formulations in order to achieve full legal status. As it stands today, American Colloid Company of Arlington Heights, Illinois is the leader in attempting a solution to the problem. In the meantime an unbelievable amount of literature has been developed by T. L. Senn of Clemson and other researchers validating the merit of humate materials and — to some extent — replication of trials.

Temperature, moisture, soil carbon, the seasons of the year, all conspire to turn humate materials into an oscillating catalyst, one never in a state of stability.

There are several firms in Florida that introduce humates into liquid nitrogens. The end product can hold nitrogen in the root zone an additional 60 to 100 days. As it is, generally, we can take different humates on the market and achieve varying results according to the materials used to liquify them.

In the main, three compounds are used to liquify humates. One is potassium hydroxide with distilled water. The purer the water, the better. Potassium hydroxide is used to drive pH up to a reading of 12 or higher. Then as much humate material as possible can be introduced into the solution. The best way to do this is to take an old milk tank with a working paddle, fill the tank with water, add the potassium hydroxide, watch the pH achieve the target level, then add the humate and let the paddle turn at a slow pace for three or four days.

Another compound that can be used is ammonium hydroxide. Still another is sodium carbonate or soda ash, $NaCO_3$. It takes about one pound of sodium carbonate to four pounds of humates. There are also available a couple of organic acids with a high pH that will push the solution to pH 12 or 13, notably Sodium Meta (silicate) brand, which is an alkalizing product. Silicate holds energy in a stronger grasp. The idea is to drive the pH of the solution high enough to permit the humate to dissolve. Many farmers

WATER CONTAMINANTS

Aluminum ammonium sulfate
Aluminum chloride solution
Aluminum potassium sulfate
Aluminum sulfate
Alum liquid
Ammonia, anhydrous
Ammonia, aqua
Ammonium silicofluoride
Ammonium sulfate
Bentonite
Bromine
Calcium carbonate
Calcium hydroxide
Calcium hyposchlorite
Calcium oxide
Calcium dioxide
Chlorinated copperas
Chlorinated lime
Chlorine
Chlorine dioxide
Copper sulfate
Disodium phosphate
Dolomitic hydrated lime

Dolomitic lime
Ferric chloride
Ferric sulfate
Fluosilicic acid
Hydrofluoric acid
Ozone
Sodium aluminate
Sodium bicarbonate
Sodium bisulfite
Sodium carbonate
Sodium chloride
Sodium fluoride
Sodium hexametaphosphate
Sodium hypochlorite
Sodium hyroxide
Sodium silicate
Sodium sulfite
Sodium thiosulfate
Sulfur dioxide
Sulfuric acid
Tetrasodium pyrophosphate
Trisodium phosphate

then use testing techniques to find one result preferable over another.

Use of the humates itself without catalyst water on sandy soil should be limited to a half pint to a quart per acre. When a half pint to a quart per acre is used, 20 gallons of water should be adequate as a carrier. If nitrogen is added — let's say ten gallons— then an additional ten gallons of water should be used.

For a conventional farm sprayer with a basic fan-type nozzle, it would take approximately 20 gallons to achieve the necessary blanket effect.

As a codicil to the above, it should be noted that household ammonia reacted with molasses will work very nicely. First, make aqua ammonia by trickling anhydrous through water, then mix molasses with it. This material, banded down the row, eliminates foxtail. Banding releases calcium which makes it uncomfortable for foxtail, a weed that will not grow in high calcium soil. Anhydrous displaces calcium, but bonded with molasses makes it more effective and does less harm to the soil.

In print, the above procedures sound simple in the extreme, and they are — except! If you ever try to liquify humates, you will likely decide that it is easier to purchase the factory-made product. The actual mixing is a breathtaking mess. First, the finely ground hu-mate material is much like powdered coal. It gets into everything and requires months to clean up.

Some of the materials are not difficult to find. Sodium carbonate, for instance, is used in most municipal water plants to soften water. Usually, it is for sale at an inexpensive price. It is hard on equipment. It comes in a white dry powder and mixes well in distilled or reverse osmosis water. It will not mix as well in hard water.

Lye should work. Lewis lye is simply sodium hydroxide, or NaOH. It contains no beneficial plant food, for which reason it should be shunned. Other products mentioned here supply ammonia and/or potassium. With lye you get a salt which you don't need. Carbon is the element to be split from the humate.

Fertility management, good water, seeds — all merit single-factor evaluation, and yet the art of quality crop production seems

POLLUTION EXAMPLES

AAtrex
Abate
Acaricide
Avenge
Banex
Banvel
Bicep
Bladex
Bravo
Fatal
Halto
acifluorfen
alachlor
alloxydim-sodium
ametryn
amine methanearsonates

amitrole
atrazine
Carbaryl
chlometoxynil
Clobber
chlorbromuron
2,4-D
chlorfenprop-methyl
DCPC
dicamba
Paraquat
dichlorprop
Endrin
dinoseb
fenuron

simpler than the sum of its parts. The *1955 Yearbook of Agriculture* put out by the USDA remains the most outstanding reference work on water for agriculture, and yet it is not all that helpful in dealing with chemical era problems.

Joe Cocannouer's *Water and the Cycle of Life* deals with the ecology of water and traces that wonderful commodity from ocean to cloud, then through the pores of the soil to inland seas and back to the ocean, but Cocannouer too wrote before toxic genetic chemicals had become an ubiquitous scourge.

Some of the contaminants encountered by farmers these days — powerful herbicides, for instance — will test even the best equipment. There are literally hundreds of pesticides entering creeks, ground water and aquifers as a consequence of our "theory period" agriculture. Even where water treatment facilities are available, they have not proved capable of removing many farm chemicals from the water. When water treatment plants remove pollution, they add dozens of chemicals to make water "safe" to drink in terms of bacteria, or to accomplish some form of mass medication. On the facing page are chemical additives approved by health officials, all of which affect and disturb the fine-tuned balance anion and cation computations call for.

In addition to the above, eleven radionuclides are found in many water supplies. Radioactivity needs minerals to hang onto. Pipeline water is particularly loaded with dissolved solids. Some manuals list over 40 pollutants routinely found in this transport system. Chlorine is particularly offensive. In the human population it is linked to heart attacks and circulatory disease. Chlorine also combines with other chemicals in water to form chloroform, a cancer causer. In the box on page 54 are a few of the hundreds of chemicals farmers use which can end up in ground water.

Farm Chemicals Handbook lists not hundreds, but thousands of these chemicals of organic synthesis.

I mentioned a herbicide earlier in this chapter. It is a difficult poison, meaning it is more difficult to deal with than Paraquat, 2,4-D, 2,4,5-T or a combination of the last two called Agent Orange. I've seen corn fields shrivel up and drop over dead in the wake of an application. It has an affinity for the clay molecule that gives it homestead status until Resurrection Day. It drops the energy field not just a little, but a lot. It seems to be a different product from lot to lot, but this is also true of fertilizers and biologicals fabricated for field application. I once did some foliar work on an alfalfa field three miles from my place, and it turned out to be a near disaster because that farmer had used the least tolerable herbicide four years earlier.

Tanks contaminated with herbicides, water sources that are polluted, both must be considered when dealing with the anatomy of fertility management.

Distilled water is possibly best, but it may not be practical when required in volume. Here are various systems of water purification.

Boiled water. Boiling water will kill bacteria, but it will actually concentrate chemicals and minerals.

Distilled water. Distilling water kills bacteria, but it leaves their bodies behind with the chemicals and minerals it removes. It handles all chemicals with specific gravity greater than water, which cancels out all but a few man-made molecules.

Rain water. The sun has distilled rain water, so it should have no germs or minerals, but it has been passed through polluted air, which is dissolved.

Filtered water. Filters cannot remove bacteria, viruses and many chemicals. What they do remove decays and contaminates all the water that passes through. Certain bone charcoal filters that back flush are an exception. Rockland U.S.A. of Tulsa is an excellent supplier of valued water systems, distillers and RO units included.

Deionized water. Deionizers effectively remove dissolved minerals, but cannot remove colloids, undissolved solids and particulates.

Reverse Osmosis. Home water pressure is too low for reverse osmosis to work. Membrane replacement is a requirement. With proper pressure, RO is one of the best purification systems.

Electrostatic. Some forms of treatment prevent accumulations of minerals in water by harnessing minerals to water for transport through the system.

When a seed is inserted in the soil, it takes on moisture and expands. It is, after all, a fertilized, matured ovule, the result of sexual reproduction in plants. The seeds of flowering plants are known as angiosperms. They are the pits, the fruits, the grains, the beans, the nuts — in short, the part of a plant a farmer sells. Carbon is what determines the capacity of a seed to take on moisture. The energy within the seed is determined by both the manganese and the payload of sugars contained therein.

There are many natural sugars in soil. Some are five carbon chain sugars, some are six chain sugars. The basis of sugar is carbon — and hydrogen and oxygen. When the seed comes up, it first shoots out a tap root regardless of how it is put into the soil. This root curls around and obeys the law of gravity.

Another anomaly of the seed is its sense of direction. If you're lost and can't figure out which way is north, simply dig up a plant and look at the root system. In the northern hemisphere, the side with the most rootlets should be north.

If there is no manganese in the seed, it will swell up and rot. Manganese has a high atomic weight, 54.9380, meaning it has more power than nutrients in the surrounding soil. This puts into play the magnetism necessary to draw nutrients into the seed to feed it and its emerging root system. When there is a shortfall for manganese, the entire fertility program has to be adjusted to create enough energy to pull more manganese.

We may never know why one lot of seeds outperforms another, same variety and specific hybrid number. We point to age and continuing respiration, and yet seeds hundreds of years old — Indian seeds and seeds stored in the pyramids of Egypt — have been sprouted and grown out. Obviously, properly balanced nutrition governs the biochemistry of the seed.

Those who are inclined to do so can easily prove out a seed's potential. I have had clients foliar feed beans commercially grown for comparison with home grown bean seeds that were sprayed in spring. The commercial beans had the proper credentials, and yet when we looked around the field there were all kinds of holes. The home grown beans that were beneficiaries of Biological Theory of Ionization treatment presented a lush, straight stand from

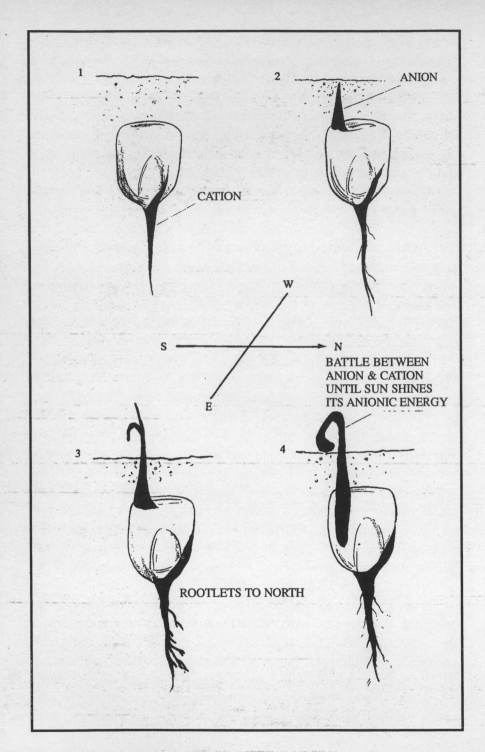

1

2 ANION

CATION

W

S ———→ N

E

BATTLE BETWEEN
ANION & CATION
UNTIL SUN SHINES
ITS ANIONIC ENERGY

3

4

ROOTLETS TO NORTH

one end of the field to the other. Not one seed seemed to fail. The commercial seed was marginal. The Biological Theory of Ionization seed exhibited potential five-fold stronger.

The final objective remains the same — anion and cation release for plant growth. The chemical compounds and fertilizers might be easier to handle mentally, but in the final analysis the fullest measure of success in crop production depends on understanding the abstractions Einstein gave to Reams, and using the arithmetic of anions and cations to calculate fertilization of crops for the growing season.

6

OF SOILS AND NUTRIENTS

Phosphate, among other things, is a catalyst, and as such it recycles. It has a function that is special, for it guides all elements into the plant except nitrogen. In other words, all elements go into the plant in phosphate form except nitrogen. Somewhere along the line there has to be a union of the phosphate atom with necessary nutrient elements for healthy plant growth. If there is a phosphate insufficiency, the plant can still uptake nutrients, but they will not be incorporated into the cell. The consequence is shrinkage. When the crop is hay, shrinkage can make the crop almost vanish. A third of the corn crop can disappear because of shrinkage. The alfalfa crop is literally annihilated when there is a phosphate shortfall. Stems will be hollow, and the difference between a hollow stem and a solid stem is the difference between half a yield and a full yield.

The last cutting on a farm I worked with had 80% solid stemmed alfalfa when we foliar fed after each harvest. The yield

THE CHEMISTRY OF NUTRIENTS

Nitrate and Ammonia Nitrogen tests indicate the nitrogen immediately available to plants, but they do not tell how much nitrogen may later be liberated from organic matter in the soil. Excess nitrogen can be harmful to plants and may leach to nearby waterways.

Phosphorus compounds in soils are slowly released to plants during the growing season and their availability is difficult to determine by chemical tests. Both acid and alkaline soils fix phosphorus in unavailable forms and annual fertilization may often be required.

Potassium available to most annual crops is indicated by the tests. However, many perennial forage crops, shrubs, and trees seem able to obtain considerable potassium from more slowly soluble forms not indicated by the tests.

Calcium status of the soil is readily revealed by the tests. Lime can correct for calcium deficiency, and will also neutralize soil acidity. Some plants grow best in acid soils, however, and excess lime may be harmful in such cases.

Magnesium tests identify soils where magnesium treatments as dolomitic lime or as sulfate of magnesia are likely to be beneficial. They also indicate whether or not magnesium is suitably balanced in proportion to potassium and calcium in the soil.

Aluminum is not required for plant growth but is associated with soil acidity and is harmful to acid-sensitive crops. Liming acid soils reduces aluminum toxicity.

Manganese is also closely related to acidity. High manganese concentrations in strongly acid soils may cause crop injury which can be decreased by liming.

Other Elements. Plants require small amounts of a number of other elements including iron, copper, zinc, sulfur and boron. Soil tests have been devised for these elements but they are affected by soil acidity much as are aluminum and manganese, and we usually may infer their availability from these tests. Sulfur is rarely deficient, while boron deficiency is usually encountered only where soils have been over-limed. Measurements of soluble salts are sometimes reported on our tests where over-fertilization or other sources of salt may have caused plant injury — *The Connecticut Agricultural Experiment Station* was greatest on the fourth cutting even though it wasn't taller — but there was no shrink.

A thinner stem may permit a larger population, but if the soil has insufficient nutrients there will not be enough energy to support the plants. If corn is healthy, tubules will be packed together all the way to the center. The center or pith of the stalk should be pearly white, not the dirty gray called gummosis.

Excess nitrogen will reveal a black layer node when the stalk is cut and put under the microscope. Such tubules often are completely blocked, much like a water pipe that is completely clogged. This is always an indication that there is not enough phosphate and calcium in relation to nitrogen.

Peppermint and spearmint do not have hollow stems if correct mineralization has been part of the fertility program. Even oats have solid stems if phosphate levels are maintained correctly to permit cell nourishment and growth. Properly nourished and nurtured, such oat stems will be more like a sturdy willow than a fragile soda fountain straw.

The research station at Bethesda, Maryland fed phosphate through the leaf in order to measure the effect on rootlets. Workers found that phosphate will travel to the roots at the rate of three feet per second. When it reaches the rootlet it forms an organic acid and soublizes fertility elements for plant uptake. But once phosphate reaches a basic level in the soil, its need is greatly reduced.

Nitrogen can carry all essential nutrients into the plant, potassium included. That is why much of agriculture grows crops with a combination of nitrogen, potassium and lots of water. This approach paints the fields deep green, but at harvest the shrink is fantastic. It reminds one of grocery store hamburger made to look superb by blending the meat with crushed ice. That same hamburger melts away in a hot skillet.

The same thing applies to livestock. It is possible to simulate growth and weight by feeding more nitrogen and potassium and keeping the phosphate level down. The gain is simply water in the cells. In a skillet or roaster, such meat shrinks and at the table it tastes like cardboard because minerals and nutrients for really good quality meat simply weren't there.

The environment around you will tell most of the story if you see what you look at. If you go through an area where all the trees have branches bushed out at the top but there are no branches down the tree, that is an indication of a phosphate deficiency or a lack of availability to the plant. If a tree is branched out all the way to the ground, that indicates a good phosphate level in the area, or perhaps that there was one.

Light green or pale green moss on a tree is sometimes an indication of iron deficiency. It could and it could not be iron deficiency. Remember, every nutrient enters a plant in a phosphate form. There simply may not have been enough phosphate to usher in the iron. So the iron may be there, albeit stalled in the plant's own horse latitudes. This is the major shortfall of leaf analysis.

Let's consider zinc. Zinc is an excellent grass promoter. But will it carry a current? Remember, anions do not conduct electricity. Sometimes the only reason a deficiency shows up is the lack of an electrolyte. When zinc, iron, copper or even magnesium are put on a plant, these applications increase the rate of ionization, and the process may create enough magnetism to charge up the growth process again. It does not necessarily mean that the problem has been corrected, but there is a response.

The only readily available tool to discern the true situation is the refractometer. Most of the time sugars go down if there is a phosphate problem, and those same sugars go up in the weeds. Most deficiencies are rarely in the trace mineral realm. They usually involve the basics — calcium, nitrogen and potassium. If these elements exhibit structural balance, the so-called minor elements are seldom a problem.

It is always difficult to design a cropping program when you do not know what a healthy crop is supposed to look like. I recall a conference in Michigan during which a farmer presented me with a Pioneer hybrid number that was completely "full kernel" with a short dent. It was solid. It not only got rid of the bigger dent, it also filled in those kernels all the way so that they were solid at the base instead of shrunk and shriveled.

If you really want to know what a crop should look like, I suggest a trip into the hills of Pennsylvania. If you can figure out some way to get an elderly Amishman to take you on a tour of his farm and share what he knows, you'll learn more than any university will ever teach. I was on one such farm some years ago. The Amishman showed me his hybrid corn. I've never seen anything like it in the upper midwest. Those stalks were unbelievable, and they didn't shrink. He showed me the animals he was feeding, which was even more impressive. He'd never purchased a bushel of corn from others in his life.

If you know what a healthy crop looks like, then you can measure your success. When you go to a grocery store to buy cauliflower or broccoli, examine the produce to see if it has a hollow area in the stem. If it has such a spot, it has a boron deficiency.

Many people have become enchanted with the term *organic*. In chemistry, anything based on carbon is considered an organic compound — and there are thousands. That's not what organic growers mean. They are simply saying that a more natural system of farming relies on the primacy of organic matter. It is the function of literacy for a person to comprehend the meaning of a term in the context of its usage. An understanding of the role of carbon in the soil in no way obliges one to accept the use of man-made molecules of chemistry with a carbon connection.

Again, carbon in the molecular structure of the seed brings water into the soil. In our opening chapter we made the point that one part carbon will hold four parts water. There are two million pounds of soil in the top six inches of an acre. A 1% organic matter soil will thus contain 20,000 pounds of carbon, and 20,000 pounds of carbon will absorb 80,000 pounds of water — or 10,000 gallons. It takes 28,000 gallons of water to cover an acre one inch deep. The problem of a three inch rain on 1% organic matter soil is at once apparent. Even a 5% organic matter soil — which is difficult to achieve under row crop conditions — would have only 100,000 pounds of carbon, and therefore a potential for holding 400,000 pounds of water, approximately 50,000 gallons, only enough capacity to absorb a two inch rain. Once a saturation point is reached, the rest of the water will run off. The soil management problem is further complicated by hardpan, which prevents water from moving down into a water dome or aquifer and forces it to run off.

With good biologically active carbon in the soil, there will still be a complement of soil air. Where there is hardpan there is no such thing as normal transportation of water up and down in the soil. A four inch hardpan loaded with salt is as tight as a dam across the river Nile. Carbon attracts moisture from the air, especially at night. If there is high humidity in the air and enough carbon in the soil, plants can get enough moisture from the air to fix a crop if there is at least 20 to 25% humidity.

Southern California was essentially desert in the early 1900s. The hills had no grass or trees. The Soil Conservation Service presided over the seeding of mountain areas with a variety of grasses. When the water wash ran off in the spring, a green layer developed and worked its way into the valley. Now when they get rain in that area, there is a basic climate change. In fact, it is possible to so manage carbon that it will change the climate of an area. It is also possible to so mismanage carbon that droughts are created. In Iowa — where they plow every possible acre from border to border — they have created droughts in areas where this phenomenon has never been heard of before.

It is difficult to get carbon in the soil to go down. Magnetism must first be created, meaning phosphate molecules must be utilized to create a condition supportive of bacteria. This means aeration — and incorporation of carbon dioxide into the soil. When air can no longer enter soil, carbon goes out as CO_2 gas. Bacteria that run into salt-rich plow pan areas die off, much as if they were cast into a salt brine tank.

The chemical symbol

Carbon
12.01115

C

6

means pure elemental carbon, a product that is difficult to achieve. We use the term *carbon*, but this expression requires a modifier. Carbon does not go in the soil as pure carbon. Generally speaking carbon is bonded with water and nitrogen to form organic acids in the soil which contain carbon. To really make soil magnetic, carbons have to be in residence to provide food for bacteria in the form of sugars.

A cornstalk has cellulose, a form of carbon. If you break it down, the breakdown products will include sugars. Bacteria can work on this cornstalk if they have a suitable environment. A mandatory component of that environment is oxygen. Another is moisture.

It is not uncommon to see cornbelt farmers put in soybeans after one corn season, then return the third year to plow up corn stalks that have been neatly embalmed. Dead soils form formaldehyde, the same stuff that's used to preserve cadavers until after the funeral. Formaldehydes are an anaerobic breakdown product. In some cases aerobes work from the top down and dilute and break out the preserved biomass. But aerobes cannot survive in formaldehyde. The remedy, again, is carbon.

Carbon, we have noted, keeps the soil from blowing, not because it is some *foo-foo dust*, but because it serves up amino acids and nitrogen — the key to stickiness, in that order, and in that order of importance. This is the soil's method of storing nitrogen from one year to the next. The conventional wisdom has farmers using a modified hydroponic system. In this view soil has little function except to prop up the crop, and maintenance of a microsystem is a luxury too costly to justify. Such a soil on injectable nitrogen is much like a drug addict. It becomes dependent on the needle arrangement.

No-till is to a large extent needle-till, forever dependent on hard chemistry. There is also a negative aspect to no-till, an inability to get the carbon to go down without air. No-till works best if the crop residue is incorporated into at least the top two inches of soil — a sort of contradiction. There has to be soil contact for microbial breakdown. There is usually as much biomass under as above the soil.

Minimum tillage, in the beginning, is better than most management systems in keeping topsoil from blowing. With residue incorporated in the top inch or two of soil, it permits enough contact for meaningful activity. Basically, organic matter is some form of plant or animal life. Mixed with the life and work of microorganisms, organic matter delivers a most valuable constituent, carbon. Carbon can also come into an active soil through the air via the agency of bacteria.

Another source is photosynthesis. The leaf takes in CO_2 from the atmosphere through its stomata. Organic matter in the soil decomposes under proper conditions, releasing carbon dioxide for plant use. Decomposing bacteria break down into humus — a point at which parent material can no longer be recognized — organic material such as corn stover. The efficiency of this process is governed by the ratio of carbon to nitrogen in the soil, which at its optimum level should be twelve parts of carbon to one part of nitrogen.

Sulfur and calcium merit consideration. Calcium, especially, becomes a problem in many areas because it does not get into the plant in the first place. Even farmers who use generous amounts of compost on their crop acres still need high-calcium lime because it just isn't there, pH notwithstanding. I have had toe-to-toe debates with folklore organic people who could conceive of no possible way lime could be in deficit. One in particular didn't believe calcium could leave his dairy farm. He forgot that people drink milk largely to get calcium. He'd been selling milk off that farm for 35 years. I offered to help him calculate how much calcium had been transported off the farm in three and a half decades.

Carey Reams talked about calciums, plural. By calciums, plural, he meant that every kind of plant had calcium in it, but always in a different organic complex. Each affects a human being differently. Calcium sulfate has a different effect on *Homo sapiens* than calcium carbonate. Calcium from alfalfa and calcium from peppermint tea are each in a different complex. As a consequence, they affect the cells of the body differently. They have a different pH and a different energy potential. These observations prompt a question over whether we should use different calcium forms on the soil. The answer is, *Yes!* But always remember a point made transparently clear in the chapter on atomic arithmetic. The higher the *atomic weight* of an element, the *less required*.

Chromium has a high atomic weight (51.996). One of the biggest problems today is getting an organic complex of chrom-ium in the diet. That nutrient is simply missing in the food crop structure. There is a lot of difference between taking chromium as a food supplement and eating it in food. All things considered, the primary high atomic weight element in the food supply is calcium. Certain foods are high in chromium, namely oysters, lobsters, and others — many of which some people say we should not eat. Nutritionists relate chromium to diabetes prevention and suggest that many insulin injections across the country could be prevented if chromium somehow made it into the food supply.

The bottom line seems to be that poor soils with less than 1% organic matter are not uncommon. Midwest prairie soils were running 10 to 12% in organic matter before the arrival of the moldboard plow. Today most of them have organic matter in the 2 to 3% range. Only a few well managed soils have a 5 to 6% index. Only rarely will 8% become an entry on a soil audit. Once intensified farming is started, most excellent soils have a tendency to back down to 5 or 6%.

When such soils have a high carbon content, the roots will travel through the soil rapidly. The best cover crop is oats or wheat. Sometimes red clover is indicated if poverty soil is to be reclaimed. Rye grass allowed to grow over eight or nine inches tall in spring will actually do more harm than good for the immediate crop year. The massive root system in the top two or three inches of soil is beyond belief unless seen. If this crop is turned in before it achieves eight or nine inches of growth, it adds nitrogen. Allowed to grow beyond that limit, the effect on the nitrogen supply will be negative. Decay will rob nitrogen from the planted crop. This is true even if the rye was planted in the fall.

There is another problem with rye. If it rains for a week, rye can grow a foot in that period of time. Instead of turning in this cover crop, you can end up looking for the tractor in all the lush growth. Nevertheless, there is good feed value in rye if it is kept short, although it may be a bit watery. Animals will likely get diarrhea, and even internal rumbling. There are several rye grasses — perennial rye, annual rye, and grain crop rye. The references here are to perennial rye, seeded at one bushel per acre.

Clovers have their own dimensions. There is a medium red clover and a mammoth variety, the mammoth being the hardy one. Red clover will grow anywhere if there is moisture.

Winter cover crops aerate the soil, but they leave very little humus after breakdown. In almost all cases, crop residues achieve a more perfect breakdown if given an assist. Liquid humates and enzymes seem to be better answers than burning straw to prevent competition for nitrogen. The animal manure that yields the most humus for the soil is the one provided by the bovine species. This is because cow and steer manure has a fiber content that is not broken down as much as, say, pig manure.

Sawdust is a form of carbon, but it can backfire. Too much is treated with unspeakable chemicals nowadays. Delivered to the farm acre, these killers can create a bad scene. I know of one Michigan farm literally put out of business for three crop years by poisoned sawdust.

Much the same is true of sewage sludge. Too many chemicals and too much plastic end up in sewage systems.

Peats have merit, but the logistics of moving the materials are staggering, and the cost often prohibitive. Packing house waste and paunch manure are excellent. The biggest single problem is that abattoirs want you to take the materials all year long. This means the farmer has to be able to accommodate the supply on a continuing basis.

Storage problems become domestic problems, with farm wives objecting to tolerating even one more fly in their houses. Cannery wastes, especially pea wastes, have a lot of nitrogen for the soil. Their value is diminished only by the fact that at least 80% of the product is water, and hauling water is generally an uneconomic procedure.

There are many materials both available and with potential, depending on scale of use. Leaves are often overlooked. At one time

I served as a consultant for a boy's home. Our objective was to fertilize without cost. Grass clippings and leaves were available. With the land mellowed and aeration going, the microbial workforce helped deliver wonderful crops. I don't know how practical this would be on a big scale. Nevertheless grass clippings and leaves supply an excellent form of nitrogen.

Dehydrated poultry litter has rated my strong recommendations in the past. Carey Reams usually suggested cage manure for broad spectrum distribution on cach acre. Unfortunately business economics intervenes too often, the cost of transporting such materials from point A to point B being too high. When I see what I can do with a liquid humate to get the necessary root stimulation, perspective starts to overrule a lot of preconceived ideas. By the time you dehydrate poultry litter — considering all the expenses involved — it has to command garden store prices. There is one final consideration. The way most organic material in a raw form is managed is not the most beneficial to the soil. Composting is still the best procedure for most such materials destined for soil application. Working in sandy areas of the country where it is arid, compost application has changed the color of the soil from light to dark brown. I once worked on one farm near Vernon Center, Texas with liquid humate. I applied a single pint to the acre along with conventional N, P and K. Within three months you could see the exact line where the humate material was spread, this from an airplane. We split one 80-acre field into two forties. We planted both the same day, and fertilized both plots exactly the same, except that one plot got the humate treatment. When wheat was harvested that year, same variety, the wheat without the humate treatment went 17 bushels to the acre, and the wheat with the humate checked out at 38 bushels per acre.

Only the right frame of thought, the right system for handling abstracts can account for such results. The arithmetic of the atoms — their anions and cations — fine tunes the system, but in the final analysis it takes a lot of living and observing to set up the right questions. Plants do not pig out for the season. They need nutrients cafeteria style, according to the curve of the cycle on which they are positioned as the season passes. There is a daily nutrient requirement for all crops, and it now becomes our function to assess it, comprehend it and guide it to crop fruition. That assessment requires thoughtful evaluation of lessons securely lodged in the literature, and it begs for a mind open to pragmatic findings as well. The entries that follow stand on the shoulders of giants and they also break trail with pioneers of the farming craft as it readies production agriculture for Century 21.

7

ORDER, NOT CHAOS, IN CHARGE

"When a fool walks through the street in the lack of understanding," quote *Ecclesiastes* 10:3, "he calls everything foolish." And if you spend your time answering and debating with fools, your time will evaporate and your results will be nil. This hostility to a new way of thinking and farming may be a burden, and it may be a blessing. Too often such people do not understand the grammar of the subject, and debating with them is much like talking to a fence post.

The Biological Theory of Ionization has to do with nutrients, not just N, P, K and calcium, but with the entire package called creation. Albert Einstein has been quoted as saying, "God doesn't play dice with the universe." It was his conception that order, not chaos, was in charge of the cosmos. That is why *rhythm* became the dominant concept in his book, *Relativity*. All life forms have their circadian rhythms. The breath of life, the heart beat that pumps 60 gallons of blood an hour in the human being, the times for nutrient intake, for sleep, for all the functions of life, are not governed by chaos, but by rhythm. The air envelope that surrounds planet earth is roughly 75% nitrogen, 23% oxygen and 2% other gases and minerals. It has been computed that some 35,000 tons of nitrogen and 10,000 tons of oxygen tower over every acre of soil. Not as clearly understood is the fact that colloidal minerals in the air routinely interact with plant growth via photosynthesis. During the last scene of all for plant residue, decomposition releases both gases and minerals into the air. Oxidation, evaporation, air movement — with or without storms — and ionization of the spheres also account for injecting vital plant nutrients into the air.

There was a time when planting according to certain phases of the moon was considered strictly folklore stuff. *The Farmer's Almanac* specialized in nailing down dates for planting much as *The Nautical Ephemerus* predicted tides a month, a year, even decades in advance of the event anywhere in the world. Only the fools mentioned in *Ecclesiastes* will deny that there is a well established connection between the phases of the moon and ionization of the air. During the time frame occupied by the new moon to the full moon sequence, there are always less ionized minerals in the air available for plant growth. Other factors figure in the availability of colloidal nutrients. Temperature figures, so does the time of the year, the locus of air over land or water as well as the type of land or water, altitude, humidity, etc.

The sun and its spots also affect the planet's ionization, and therefore the availability of trace nutrients plants routinely take from the air. I have mentioned before that plants take approximately 80% of their energy requirements from the air, sunlight not being the least important. The rest of the energy needs come from soil and water.

Arbitrarily, we fence out most of these considerations when we discuss a soils program and recommend fertilizers and ratios necessary to satisfy electromagnetic fields and the exchange between anion and anion, anion and cation, and cation and cation. It is the function of our tests and our fertilization program to expand electromagnetic fields because strong fields hold nutrients in place. Moreover, the use of computed ionization for materials applied has the dual effect of keeping soil systems cooler in summer and warmer in winter, the ideal being a constant 70 F. temperature.

The critical point here is that proper fertility management affects the air above crop land and governs interaction between plants and air, and the nutrients forever floating in the air.

Experience as well as dead reckoning has provided guidelines for proper soil maintenance. Target figures say that the acre needs 2,000 pounds of calcium, and P_2O_5 should be moving toward the 400 pound level, with 200 pounds of K_2O being appropriate for excellent production. These payloads are computed on a water soluble basis. Nitrate nitrogen and ammonium nitrogen suggest a 40 pound inventory for each. There are other data for this approach: 40 pounds iron; two pounds manganese; five pounds copper. Added together, these data come to somewhere around 2,600 to 2,800 pounds of active plant food.

These fertilizers cannot function without burning and dehydrating the crop unless organic matter is near 5%. This rule holds for both dry land and irrigated acres.

The goal is TDN, total digestive nutrients — nitrogen, calcium, phosphorus, potassium. It is the function of carbon to keep these nutrients separated by enough space to confer on them the status of *complexes*, and keep them from becoming salts.

A point in question — anhydrous ammonia that is not combined with a carbon. It is very reactive and it will grab other nutrients, complexing them. It will reach into the air and invade the soil's water supply. Introduced into low organic matter soil, it will burn it up, quickly annihilating that bank account. It will bond to the available carbon supply. This nitrogen bonded to carbon in the presence of a water supply will yield an amino acid. Such an amino acid will not move up and down in the soil because it is sticky. Complex carbon and nitrogen molecules reject a potassium bond, and failure of this bonding mechanism results in a salt. The carbons have to hold the several nitrogens separated. When carbon — or generally the absence thereof — fails to maintain the peace between nitrogen and potassium, the fires of war erupt, and in the process plants get burned by the formation of resultant salts.

The principle can be best illustrated by considering grape juice, perhaps bananas. The brain has a high requirement for potassium. Heavy brain work — writing, lecturing, making computations or answering the demands of scientific investigation, all burn potassium in breath-taking amounts. Yet if I take the water and carbon out of grape juice, I end up with a salt. A salt brine solution will draw the water out of a plant and condense it. This will eventually burn plant roots. When too much salt ends up in the root zone, it is often possible to pull a plant out of the soil and slide off the rootbark. This tells you that salt concentrations are too heavy and carbon is deficient.

A chelate is an element that carries an extra electron. Iron chelates are simply iron plus an amino acid. Normally iron has a strong positive charge, but when bonded to an amino acid, the resultant compound has a slightly negative charge. This makes for easy transport into a plant.

Chelates are equally valid in animal husbandry. They permit transport of iron across intestinal walls much like a soft shoe entry, sans side effects common with a straight iron compound. This permits rapid correction of a disease condition if extreme care is used. The law-of-the-little-bit applies here: a little bit goes a long way.

A catalyst is any substance that can join two or more elements or compounds together without itself becoming a part of the union. A catalyst is much like a wrench used to tighten a bolt, if such a homey illustration can be permitted. Once the job is done, it apes caducity — that quality of being transitory or perishable. Under most modern agronomy systems, unused phosphates perish, so to speak, by being locked up and made unavailable, a not too proper role for a catalyst. The phosphate element also figures in cell construction.

The world's conventional agriculture will have to face revision almost immediately. The wrench should be good for crop after crop, yet each year conventional agriculture throws the wrench away by embedding it in the concrete of unavailability. The system is wasteful because it rarely uses 15% of the phosphate supplied. This state of affairs continues, even though phosphate-rich

pNa METER

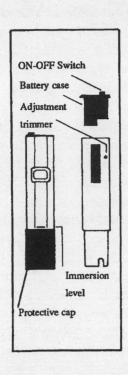
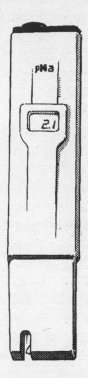

The pNa tester utilizes a sodium ion selective glass electrode to determine the acivity of free sodium in solution.

$$pNa = \log a_{Na}$$

In dilute solutions, the activity coefficient is nearly 1 and in such situations the pNa works as a good indicator of the sodium ion concentration.

A double junction reference can be used to ensure a highly stable reading. If the calibration has drifted, recalibration can be achieved with a solution of known concentration by using the calibration trimmer.

islands in the Pacific are being mined down to ocean level. World-wide, this phosphate waste — made necessary by dead soils, a loss of soil life — by improper use of salt fertilizers and toxic rescue chemistry, in short — has taken supplies and potential supplies to a dangerous level.

Water is the main catalyst and phosphate is second. Phosphate joins the many elements needed to build and energize a food cell for human consumption. Only the plant food that is made soluble in water is suitable for plant uptake. This, of course, is the theory behind water soluble fertilizers: the plants use them immediately with stronger magnetism within the plant an observed and measured result. Magnetic strength enables a plant to pull from the soil nutrients it might be otherwise forced to forego. These fertilizers are not poisons and should not be compared to chemicals of organic synthesis, as is common among organic folk. But they become available outside the organization of nature. That is why they are both dangerous and dramatic. It takes organic matter to buffer their use, its presence or absence dictating not only the amount per application, but also the timing of applications as well.

Trillions, maybe even quadrillions of unpaid microbial workers are the key to the organic matter, humus, carbon, fertilizer equation, with the much maligned anerobes leading the way. They convert crop stover into humus. pH of the soil governs their identity and function, with the pseudomonas group operative in a pH range of from 4 to 5.8, and *Nitrous pseudomonas* taking nitrogen from the air when the pH ranges between 5.8 and 7.0. Higher on the scale are soils with a pH 7 to 9 reading and still a different inventory of microorganisms with names still being evolved in the manuals.

With *Nitrous pseudomonas* on the job, and a proper carbon complex in the soil, nature's own prestidigitation takes over. Nitrogen and water are drawn from the air by magic — there is not another word for it, because a special sequence of events takes place.

There is an area in Arizona where an increase of carbon in the soil has brought on an increase of rainfall. The combination foundation and trigger mechanism has been enzymes and liquid humates. In the living desert there are a special anaerobic bacteria you really don't want. They destroy organic materials. They convert them into aldehydes, the primary one being formaldehyde. There is a whole muster of these oxygenphobes — to suggest a term — which do more damage to the soil than a labor of moles. Once these spoiler bacteria do their thing they literally embalm crop residues and set up an environment in which aerobes won't grow. This means organic matter in the soil turns to ash, then salt, the precursor of hardpan.

Hardpan is to soils what a pace of slow asses is to traffic on a mountain trail. It crushes progress to death. At about the 16 inch level in the cornbelt, at a six inch depth in the high plains, at a 24 inch depth in some alluvial areas, a layer four to eight inches thick develops, hard as a granite plug, and containing enough salts to keep it that way. Children sometimes synthesize the consistency of hardpan by failing to put away their play dough. Even a first grader can make the stuff — with a little flour, water and to make it harden, common table salt, or sodium chloride.

It takes no sleuth to identify a soil with hardpan. A small sledge hammer and a metal fence post will do. In a good healthy soil the fence post probe will slide right down, perhaps three feet. I said *in a good healthy soil*. Now take the average soil. Most of the time the fence post probe will go down 12 to 16 inches and stop. Likely as not the hammer will sing into the air if the strike is not accurate to a marked degree. The first impulse is to figure you've hit a slab of stone or buried concrete. If the hammer manages to drive on down through these several inches, then the post will continue its journey toward the center of the earth. The steel fence post is a marvel for identifying the hardpan, and useless in doing much about it. Obviously there is no conduit for water to come up through the subsoil at night, when atmospheric pressure causes the soil to try to take a deep breath. As a consequence, crops become extremely dependent on regular rainfall.

The fertility program I design and teach will enable any willing farmer to lift the pall that often settles over a soil's potential. Starting the movement of subsoil moisture through the hardpan area is both a challenge and a reward. As salts dissolve and dilute out, it becomes the start of eventually getting rid of the hardpan.

Like bad "good news," "bad news" jokes, there is indeed a bit of good news associated with the hardpan layer. It prevents chemical contamination from going down through the salt layer. Toxic runoff is no blessing, but it may be preferred to having a trap door into the resident aquifer. In the sandy soil areas of Wisconsin and Nebraska, with or without hardpan areas, the destination for toxicity is always the aquifer.

Water has the same regulator as that sovereign over salt formation and salts, namely carbon. Specifically, organically completed carbon holds up to four times its weight in water, and a soil without a fair carbon complex is too soon overloaded with water that leaches out minerals. This same faltering carbon profile in the soil facilitates aquifer and water dome contamination with chemicals of organic synthesis and salt fertilizers.

Withal, it is the nature of water to carry nature's baggage and man's mistakes. Water is the delicate requirement for soil, plant, animal and man. Without it plants won't grow and *Homo sapiens* won't drum a finger. It either cleanses or pollutes every cell it touches. The male sex is 80% water; females, 70%. We function because blood — which is 90% water — passes through the kidneys at the rate of 1,000 quarts every 24 hours, taking away toxins that will kill if given a chance. The only thing more necessary is oxygen.

Managed or mismanaged by modern technology, raw water is generally full of bacteria and viruses. It is also loaded with industrial wastes, but agricultural wastes are starting to dominate. The chemical trade press admits to having some 80,000 toxic chemicals on the market in the U.S., with 1,300 new preparations coming on-line every year, all of them carcinogenic, mutagenic or teratogenic by structure.

A human being should drink a quart of pure water a day. And therein lies the rub. Merely drinking pure water is not enough. The question has been posed by the medical literature — given the choice of drinking two liters of water contaminated with a

PENETROMETER

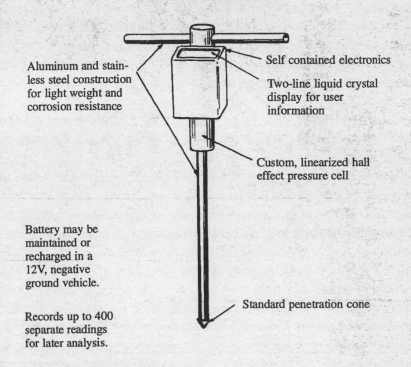

Aluminum and stainless steel construction for light weight and corrosion resistance

Self contained electronics

Two-line liquid crystal display for user information

Custom, linearized hall effect pressure cell

Battery may be maintained or recharged in a 12V, negative ground vehicle.

Records up to 400 separate readings for later analysis.

Data may be analyzed via:
1. Data terminal
2. Personal computer running communications program
3. Personal computer running analysis program

Standard penetration cone

Penetration pressure recorded at each 1/4 inch of penetration up to 18 inches.

Reading may be terminated at any depth or restarted if hard object (*e.g.*, a rock) is encountered.

volatile chemical at seven ppb or showering for 15 minutes in the same water, which would be the wisest choice. The answer is to drink it, not shower in it, because four times more chemicals would enter the body from taking a bath than from taking a drink, according to some published literature.

Chlorine, fluorine, dissolved solids, toxic chemicals — all condemn the water systems, just as the water systems condemn people. In the news people wondered why so many children in the Port Arthur, New York area were being born without legs and arms. They need wonder no more. Trichloroethylene was found in the water at a level 1,100 times greater than the safe level EPA conjured up. In fact, there is no safe level and no tolerance level.

In any case, drinking a quart of pure water will bring on uplifting benedictions of change to the body chemistry. The proof is easily established. A sample of urine in a conductivity meter will measure the salt concentration. A high salt concentration offends human cells just as a salt overload offends plants. With or without instrumentation, too little water intake will result in cloudy urine.

For now it is enough to make the air, soil, water connection with photosynthesis. Osmosis is the process of sap going up. Photosynthesis is the process of converting sap to sugar, and this includes bringing the sap down to the pons and redistributing the sugar to the rest of the plant. The pons is a sort of brain center for the plant, and it is usually right at the ground level. In photosynthesis there is one limiting factor in putting sugars into plants, namely phosphate, the same catalyst phosphate mentioned earlier. When sugars will not come up in the plant, then phosphates are not getting into the plant. Something, somewhere in the soil structure or in the soil chemistry has run amok. During the daytime when heat from the sun shines on a leaf, water molecules within the leaf expand.

Expansion serves up two basic ingredients for sugar — oxygen and hydrogen. Needed now to form sugar is carbon, the last of the necessary triumvirate. In the leaf, phosphate causes enough of expansion in the water molecule to permit entry of CO_2, carbon dioxide. This takes place during the day, during the expansion phase. It is phosphate that causes the hydrogen and the oxygen from the water molecule to bond with carbon dioxide in the air. Carbon dioxide from the air is taken in at sites on the underside of the leaf called stomata. Magnified, they look like small cavities. Miniature hairlike projections rise from around each orifice like antennas, which they are. The length of those antennas determines which elements the plant can fetch most effectively from the air surrounding it. Carbon dioxide is the primary nutrient, but there are others — nitrogen, magnesium, potassium — provided the plant has enough magnetism to lure these elements into those tiny craters. The more heat the leaf collects, the more water expands and the more potential for sugar formation if the catalyst phosphate is present. In the evening the leaf contracts and captures the carbon. The phosphate is more or less spit out. Once the catalyst phosphate is relieved of its duty, the carbon, hydrogen and oxygen molecule, $C_{12}H_{22}O_{11}$, or some such combination results.

The mysteries and contradictions of heat and cold, and resultant chemical reactions, can be illustrated — if not explained — by the heating and cooling of water. Heat water and it expands. Cool water and it contracts, but when it gets cold enough it expands again.

My students often have questions, which is fine. But what pleases me most is their willingness to stay on for the answers. My patience runs thin at times when someone wants me to back away from the lessons the Bible and Christianity have to offer. And yet I feel sorry for those who cannot make the connection. Many times in life I have to say to myself, *There but for the grace of God . . .*

I mentioned the pons area of plants a little earlier. There is definitely such an area. If you buy pecan trees you will discern the fact that they have a forked root system. If this is missing, the pons will be missing, and the tree will grow a year or two and die.

I was hired by a man from southern Missouri a few seasons ago. He wanted a seminar in private. I was very busy at the time, so after the fourth phone call I decided to fix the situation. I decided that if I ran the fee high enough, I'd take care of the problem. I called the gentleman back and told him my fee. He responded, "Fantastic, when can I come?" I begged off to the point that I knew very little about pecans. He persisted. "I've heard that you have the information I need," he said. He wheeled into my yard one day. When I looked out the window I could have crawled under a rock. There was a gentleman who was twenty-eight years old. He had a wheelchair hanging on one side of his Toyota. He was a paraplegic. I had a hard time all day long, but I sat down and gave him the best I knew. When I got done, there was no way I was going to accept money from him. I told him that I really didn't want to do this and that I had put the fee high enough so he wouldn't come. He said, *No, you've more than earned your keep.* He gave me the check with tears in his eyes and said, *I've been looking for this for the last five years, and what you told me today is worth ten times this much.* He had been purchasing pecans and he was out there in his wheelchair managing the grove. Every year over 80% of the trees would die off. I explained to him about the pons and the forked root on the pecan tree. When he heard this, he said he knew exactly what they had been doing to him all along. The nursery where he had been buying knew it. They knew that without the pons, root stocks would languish and die. This farmer was buying up to 3,000 of these plants at a dollar each. He concluded that he would grow pecans from seed if I would show him how. A little later I received a letter telling me about 3,000 babies on the front porch. He had learned how to set them out in a kelp mixture with lime. He rolled the root mass in burlap bags and set them in rows with the objective of selling them as seedlings. He concluded that if there was so little good stock and honesty in the market, he'd start delivering both.

In science as in the affairs of men, there is this conceit that the laws of nature are not the same as the laws of God. Chaos is believed to be in charge. What I offer in the text of this message is some understanding of the soil-plant relationship from the other, not a chaos point of view.

As a practical matter we have to make pragmatic observations. We have to scrutinize and even use some eco-agriculture ideas and deal with symptoms, albeit with substances that are more acceptable to rational people. Frankly, we can't "have done" with all trial-and-error. But our last line has to be the one that opened this chapter — "When a fool walks through the street in the lack of understanding, he calls everything foolish."

8

ALL THINGS CONSIDERED

The Great Pyramid on the plain of Giza, near Cairo, is said to have locked in its dimensions a sequence of numbers generally associated with Leonardo of Pisa, who was nicknamed Fibonacci. He was an Italian merchant who worked within one of the Arabic infiltration routes during a part of the twelfth and thirteenth centuries. In 1202 he published a textbook using algebra and Hindu-Arabic numerals. In it he proclaimed the sequence that bears his name,

1, 1, 2, 3, 5, 8, 13, 21, 34, 55, 89. . .

Each number is the sum of the two preceding ones. The growth of pineapple cells, the heredity effects of certain human matings, and countless other patterns engineered by nature's God argue further that order, not chaos, presides in the universe. A *Fibonacci Quarterly* is published in the United States because the fertility of the sequence seems as endless as creation itself.

These few thoughts come to mind because an understanding of agriculture requires us to wonder first, then attempt to understand exactly what governs the speed of a plant's growth. Pragmatically we observe that a plant will grow as fast as we can force the nutrient into it until it starts to produce seed. At this point time becomes a factor, and cell production cannot be pushed beyond that point. Unfortunately, we have more questions than answers.

I have experimented with string beans at my farm in Minnesota. I've pushed plants to an 18 inch height by forcing nutrients into them. Oddly enough, getting super growth seems to cancel out a seed set. At one time it took Florida growers 70 to 80 days to grow a seed bearing bean crop, but now they are down to between 30 and 40 days.

Carey Reams at one time used uranium ore of a certain kind to increase the rate of ionization and cut down the growing time of a crop. If you want to increase ionization in your garden and get faster growth of vegetables, a trip to the local machine shop may be in order. They have neat iron filings you can spread around your garden. I suggest you clean them first to get rid of cutting oils that can load toxicity into your soil and cancel out the benefits of this new found trap for the equator to pole magnetic flow.

The final benediction for rapid growth is greater quality. This profound principle brings back into focus our initial lesson, that anions cause growth, and cations cause seed production or fruit. Nitrogen in nitrate form is the author of growth — rapid growth. Nitrogen in ammonium form causes seed and fruit production. There is a dilemma in all this. The anion-cation complex can be switched back and forth by the type of fertilizer used in the soil, and by the weather. Therein lies the problem of maintaining the right type of nitrogen for the type of growth sought at a given time. The Creator has decreed that a natural switch take place precisely when production crops transfer their attention to growing the plant part the farmer usually sells. But as we raise specialized crops from which we want only the leaf or forage instead of the seed, control for maximum production — seed or foliage — is something devoutly to be wished.

All elements in a molecular structure are given the same size when at the same temperature and pressure. There are two types of colloids. One of them is a measurement of size. Usually, a colloidal substance is finer than talcum powder. Poured from a container, it runs much like water. Second, there is an organic chelated compound that contains all the elements in the solar system, save one. This is soft rock phosphate. Crops can be grown in soft rock phosphate if the missing link, nitrogen, is added.

The soft rock phosphate colloid is a byproduct of hard rock phosphate mining. In the hard rock phosphate mining process, a fine powder is produced. It is separated from the parent material by wash water which carries it into huge settling ponds. The mining operations seldom realize that the byproduct — the colloidal soft rock phosphate — is really more valuable to sound agriculture than the hard rock served up to the trades for acid treatment. I have seen such a settling basin 65 acres in size, some 50 to 60 feet deep. Bowing to the conventional wisdom of agriculture, these strip mine operators layer soil over this valuable material the way iron miners sometimes bury the valuable material called taconite. They spend million to build settling ponds, then bury the colloidal product rather than sell it into production agriculture.

More recently I have gained access to some of these deposits for use in California and Arizona grape arbors. Application of this soft rock phosphate has resulted in the harvest of quality grapes with the highest sugar content ever seen in both areas.

All things considered, it is the failure to comprehend pH that undermines proper harnessing of the ion flow. Most glossaries define pH as a measure of acidity, pH 7 being neutral. Higher numbers mean alkalinity, lower numbers mean acidity. Our definition will not suggest sour soil and the addition of calcium when pH is under 7.

Magnesium, pound for pound, can raise the pH up to 1.4 times higher than calcium. A soil high in magnesium and low in calcium can test above 6.5, but will be entirely inadequate for the growth of alfalfa, for the growth of legume bacteria, and above all, for maintenance of an environment necessary to decay organic crop residues into humus.

It is more essential to manage the factors that construct the soil pH as determined by present soil tests than it is to be concerned with the calculated amount of N, P and K. Why?

Here are a few notes quoted from C.J. Fenzau, writing in *An Acres U.S.A. Desk Reference*, volume one, page 95.

The conversion and availability of mineral elements are related to and regulated by the system of decay in the soil. The proper decay processes are initiated and determined by usually full levels of calcium and reasonable levels of magnesium.

A balanced equilibrium of calcium and magnesium creates a soil environment for bacteria and fungus activity for the proper decay of organic residues into CO_2, carbonic acid and a host of many weak and mild organic acids, all so necessary to convert and release mineral elements in the soil system.

An unbalanced equilibrium of calcium and magnesium permits organic residues to decay into alcohol, a sterilant to bacteria; and into formaldehyde, a preservative of cell tissue. The symptoms of this improper decay system can be observed when previous year's stalks are plowed back up just as shiny and fresh as they were when turned down.

Under these conditions, larger and increasing amounts of nitrogen and fertilizer minerals will be required just to maintain normal crop yields. The soil system is not complementary to release the minerals nor the other soil essentials for optimum growth.

And remember also that large applications of nitrogen consume larger amounts of calcium as well as "burn-up" crop residues and humus. You can get increased yields for a few years from this stored-up wealth of humus, but eventually you will have to account for this withdrawal.

Without an active organic matter system in the soil you cannot grow any crop at all, no matter how much N, P and K you add. The soil is a living complex system that not only holds the twelve necessary minerals needed for plant life, but also is the factory that produces carbon dioxide, digests lignin into humus, provides nutrition and energy for desirable bacterial and soil animal life, and is the container for both water and air. Can you germinate and grow a plant in subsoil material without humus action present?

In the absence of a system of organic matter management it will soon be essential to consider buying dry ice (as a source of CO_2) and propane gas (as a source of additional carbon). Both are sad excuses for our ignorance and mismanagement of calcium and magnesium and at the same time a dear price to be paid for our continued disregard of this important and vital soil equilibrium.

A soil program with managed levels of calcium and magnesium will allow the nature of the soil to function complementary to the process of life and in many cases actually diminish or even eliminate the causes of many problems we can't explain or do much about ourselves. Some of these benefits and effects are:

1. More efficient photosynthesis.

2. Maximum use of heat-degree days — a natural time clock of the life system of plants.

3. Create and maintain root and stem capacities for optimum use of sunlight energy and thermal efficiency by the leaves of the plant.

4. Thereby the plants can use water, CO_2, nitrogen and mineral nutrients with greater efficiency.

5. A healthy and normal functioning plant can maintain an adequate hormone and enzyme system so vital to resistance to insect and disease hazards.

6. A balanced soil equilibrium will regulate and manage the quality and availability of all mineral elements needed by growing plants.

7. Excesses of minerals during the early growth stages often plug up the vessels of the stem and are the frequent cause of early death. A dead plant system cannot mature or ripen itself.

8. An excess of magnesium as well as nitrogen in the soil initiates the processes which prevent the crop from growing dry and nutritionally ripe which is a major goal of every farmer.

9. Carbon dioxide availability is more important to high-yield potentials than nitrogen or any mineral element. The supply in the atmosphere could sustain life for not much more than 30 days, and depends on the soil system and its effect on soil structure and tilth, the processes are retarded and inefficient.

10. Clay soils high in magnesium and low in calcium cement together tightly, are subject to compaction and clodding, crust over easily and prevent the insoak of water and the recovery of capillary water during the dry periods of the season.

11. Soils in such poor tilth and structure increase the effects of the many weather hazards that annually impair normal plant growth. With a managed calcium-magnesium equilibrium we would not have to lean on the many poor excuses of weather we use every year to explain away our ignorance and personal mistakes. We cannot do much about the weather, but we can use the experiences of our mistakes to create a more integrated soil management system that would tend to diminish or even eliminate the many variable hazards of the weather.

pH, then, is a measure of resistance. And pH 7 says there is equal resistance between cations and anions. At pH 7 synchronization has been achieved. pH, technically, is the mathematical symbol for potential of the hydrogen ion. There is a shortfall in the value of this equation because it does not tell the pounds per acre. A pH 7 reading can be had without one single pound of calcium per acre. Pure white sand has a pH 7 reading and there's not an ounce of calcium in its construction. Such a test presented to the average college classroom would get a "no lime required" opinion.

This is not the only contradiction farmers have to face. Crops need most of their annual energy requirement at the end of the growing season. But they get most of it at the beginning. The energy curve should start at, say, 50 and work toward 400 because plants feed cafeteria style, a bit at a time. They do not "pig out" at planting time for the rest of the season. In a way, farmers know this, but they proceed anyway to load the plants with the season's requirements — starting with 600 or 800 on the curve, with a vain hope of having the energy needed as harvest draws near. As a consequence, a considerable amount of yield is lost. At precisely the time when the last nutrient flush is indicated, many growers are on their family vacations.

A while back I was working with an oats producer who did not go on vacation as the critical harvest hour approached. Approximately two weeks before harvest, we were still delivering foliar nutrients to the crop. In those last two weeks we added between 20 and 25 bushels of oats per acre because we sprayed before the energy needs dropped below the critical point. I observed that after we sprayed a second time, we had new heads forming below the "first growth" heads. The university texts say the oat crop is not exacting in its soil conditions and seedbed requirements, and that the crop is not sensitive to soil acidity. If the farmer is willing to settle for a mediocre crop, these things may be true. From my chair it appears that a maximum anion-anion, cation-cation, anion-cation exchange must be kept operative down to the wire, and this means energy all the way.

In most soils, at peak a conductivity meter will read somewhere in the 400 range if things are balanced. A lot of times the reading will be in the 800 to 1,000 range. That reading is merely saying all is well with the soil and the crop. There is a second subliminal message. . . *you can* force *a crop and a high yield, but it's expensive doing it that way.*

When lime and Calphos are incorporated into the program, they will stabilize out and follow a pretty good pattern and structure a fairly consistent curve. When calcium is not being mobilized properly, the conductivity reading will exhibit erratic patterns during the growing season, but if you have adequate calcium paired with adequate carbon to stabilize the soil, that problem will not arrive. Routine monitoring of the growing season with a conductivity meter has served to bless some of the newer water soluble products on the market. When the crop is running short, these products make it possible to add a few more pounds per acre. In many cases such cafeteria style feeding can add 15 to 20 bushels per acre to the corn crop. I've done it, and it works. If, indeed, there is a shortage of available calcium and carbon in the soil, a good energy pattern that will stay stable can be established with the solubles. In other words, if you check the crop in the morning and check it again in the evening and you do this at the same time each day, you will discern fluctuation instead of maintenance of a normal steady level all the time. The crop will grow fast for a while, then it will stop and start fast again. Calcium is the main buffer for resistance in the soil. Calcium is also the major anionic element in the soil for crop production. Nearly all the other fertilizer elements you add to the soil are positively charged. If you do not have an adequate level of calcium and you add a positively charged element, an erratic spurt in growth will follow, and maintenance of a stable soil condition will prove impossible.

The key to fertility management might be compared to building a baby formula — in the beginning, anyway. The baby does not need much formula, but the right amount of everything is necessary if good health is to be the objective. In other words, you wouldn't feed junk food to a baby. There are two junk foods as far as agriculture is concerned — potassium chloride and anhydrous ammonia. These two substances are not needed, and when used they clog up cell metabolism and offend the quality requirement as expressed by specific gravity. Top quality seed corn should weigh at least 60 pounds per bushel. The standard base at

pH METER

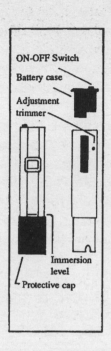

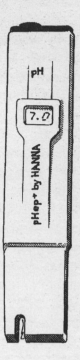

ON-OFF Switch

Battery case

Adjustment trimmer

Immersion level

Protective cap

pH

7.0

pHep+ by HANNA

This handy and easy-to-use hi-tech instrument does the work of more than 300 rolls of indicator paper. With a comfortable reading display and an accuracy of 0.2 pH, pHep is far superior to any type of indicator papers of chemical testers. Its use on-scene for monitoring water and soil conditions is indicated for mainline farming, Century 21.

an elevator is 56 pounds per bushel. This will be likely lowered to 54 pounds because that's all most farmers can now deliver based on the technology they use.

Oats should weigh 60 pounds per bushel, but they don't because 32-pound oats is what most farmers have the capability of growing. Farmers in Belgium and the Netherlands grow 60-pound oats on reclaimed ocean floor. It may be that a rich distillation of minerals in old ocean floors is accounting for heavy oats. Whatever the explanation, there can be no doubt that size does not mean weight and quality.

There is one awesome disadvantage to the use of water soluble fertilizer materials. They will leave the soil with the first rain. But when we deal with our carbon requirement, we can hold these fertility elements in the soil for a much longer period of time. The lower the analysis for a final application, the better. The bottom-line translation of all this is that to move a fertilizer from point A to point B, a concentrated form is indicated, but to get good distribution, a dilution system is required.

It takes 20 gallons of water to achieve maximum coverage, using a conventional farm sprayer. Equally as important as full knowledge of the distribution mechanism is an understanding of how the materials act. For instance, use of 18-46-0 in a high calcium soil will result in a rapid tie-up into tricalcium phosphate if the soil has low organic matter and low biological activity.

There are some eight major plant nutrients that require our attention because they are needed most by weight and volume.

Phosphorus is never in an elemental form in the soil. It is, in fact, a *phosphate compound*. A single molecule of phosphate packs a lot of energy. If oxygen is added, the Milhouse computation comes to 108,250 units, based on an average.

Two types of nitrogen also rate analysis via computation. One is an anionic nitrogen, the other is cationic nitrogen. With the seed production role of cationic nitrogen firmly in mind, it is nevertheless necessary to point out that manganese is even a more powerful governor of seed production. Cationic energy is at least two and one-half times more powerful than anionic energy. Ammonial nitrogen will switch to the nitrate form after it is applied to the

PERIODIC TABLE OF THE ELEMENTS ACCORDING TO THE BIOLOGICAL THEORY OF IONIZATION

ELEMENT	SYMBOL	ATOMIC WEIGHT	MILHOUSE UNITS
Hydrogen	H+	1	1000
Helium	He+	4	3250
Lithium	Li+	7	5500
Beryllium	Be+	9	7000
Boron	B+	11	8500
Carbon	C+	12	9250
Nitrogen	N-	14	4250
Nitrogen	N+	14	10750
Oxygen	O+	16	12250
Fluorine	F+	19	14500
Neon	Ne+	20	15250
Sodium	Na+	23	17500
Magnesium	Mg+	24	18250
Aluminum	Al+	30	22750
Silicon	Si+	28	21250
Phosphorus	P+	31	23500
Sulfur	S+	32	24250
Chlorine	Cl-	35	9500
Potassium	K-	39	10500
Calcium	Ca-	40	10750
Argon	Ar+	40	30250
Scandium	Sc+	45	34000
Titanium	Ti+	48	36250
Vanadium	V+	51	38500
Chromium	Cr+	52	39250
Manganese	Mn+	55	41500
Iron	Fe+	56	42250
Cobalt	Co+	59	44500
Nickel	Ni+	59	44500
Copper	Cu+	64	48250
Zinc	Zn+	65	49000

ELEMENT	SYMBOL	ATOMIC WEIGHT	MILHOUSE UNITS
Gallium	Ga+	73	55000
Arsenic	As+	75	56500
Selenium	Se+	79	59500
Bromine	Br+	80	60250
Krypton	Kr+	84	63250
Rubidium	Rb+	85	64000
Strontium	Sr+	88	66250
Yttrium	Y+	89	67000
Zirconium	Zr+	91	68500
Niobium	Nb+	93	70000
Molybdenum	Mo+	96	72250
Technetium	Tc+	98	73750
Ruthenium	Ru+	103	76000
Palladium	Pd+	106	79750
Silver	Ag+	108	81250
Cadmium	Cd+	112	84250
Indium	In+	115	86500
Tin	Sn+	119	89500
Antimony	Sb+	122	91750
Iodine	I+	127	95500
Tellurium	Te+	128	96250
Xenon	Xe+	131	98500
Cesium	Cs+	133	100000
Barium	Ba+	137	103000
Lanthanum	La+	139	104500
Hafnium	Hf+	178	133750
Tantalum	Ta+	181	136000
Tungsten	W+	184	138250
Rhenium	Re+	186	139750
Osmium	Os+	190	142750
Iridium	Ir+	192	144250
Platinum	Pt+	195	146500
Gold	Au+	197	148000
Mercury	Hg+	201	151000
Thallium	Tl+	204	153250
Lead	Pb+	207	155500
Bismuth	Bi+	209	157000

ELEMENT	SYMBOL	ATOMIC WEIGHT	MILHOUSE UNITS
Polonium	Po+	210	157750
Astatine	At+	210	157750
Radon	Rn+	222	166750
Francium	Fr+	223	167500
Radium	Ra+	226	169750
Actinium	Ac+	227	170500
Cerium	Ce+	140	105250
Praseodymium	Pr+	141	106000
Neodymium	Nd+	144	108250
Promethium	Pm+	147	110500
Samarium	Sm+	150	112750
Europium	Eu+	152	114250
Gadolinium	Gd+	157	118000
Terbium	Tb+	159	119500
Dysprosium	Dy+	163	122500
Holmium	Ho+	165	124000
Erbium	Er+	167	125500
Thulium	Tm+	169	127000
Ytterbium	Yb+	173	130000
Lutetium	Lu+	175	131500
Thorium	Th+	232	174250
Protoactinium	Pa+	231	173500
Uranium	U+	238	178750
Neptunium	Np+	237	178000
Plutonium	Pu+	242	181750
Americium	Am+	243	182500
Curium	Cm+	247	185500
Berkelium	Bk+	247	185500
Californium	Cf+	249	187000
Fermium	Fm+	253	190000
Einsteinium	Es+	254	190750
Mendelevium	Md+	256	192250

soil if there is enough calcium to engineer the switch. That is why it is possible to stop in its tracks an alfalfa crop short on calcium by applying ammonium sulfate, allowing it to go to seed. But if there is enough calcium, the same application of ammonium sulfates will cause a great increase in growth.

The rule of the antenna applies. If you're going to draw in a radio station, it takes more power to draw one than the other. Much the same is true of a crop. The stronger the drawing power of a plant's antenna, the faster the crop will grow. Calcium will increase the power of the antenna, ergo an increase in growth. The more calcium transported into the plant, the greater its ability to take nutrients out of the air, chiefly carbon dioxide, nitrogen, potassium and magnesium. The way conventional agriculture is managed nowadays, the farmer is fortunate to get even a little CO_2 uptake. Other elements are not taken up at all, and for this reason the average grower has to add more and more fertilizers.

Chlorine is an anionic substance, meaning that from an energy point of view it carries a negative charge. Potassium also carries a negative charge. Too much of agriculture has turned to potassium chloride in place of lime to get a crop. Trade practices are often uneconomic practices to someone. This is certainly the case with potassium chloride, the beneficiary of university hard sell and the author of industry profits. High calcium lime would be cheaper, but limestone quarries seldom earmark money for grants.

Again, all things considered, it is correct to say that nature will always follow the line of least resistance. Roots can be made to travel wherever you want them to go. In compacted soil, roots may not travel as fast and as well, but they will follow the line of least resistance laid out for them by compost or other nutrients. When you construct a trail that makes it easier for roots to travel, chances are improved for picking up all the necessary nutrients. Nutrients held in escrow in a little ball do not lure roots half as well as the same nutrient load with vastly increased surface sites. There is another rule that must be held up for consideration: the greater the density of soil without humus, the higher the specific gravity. The translation to this esoteric language is simple in the extreme. When you take the carbon out of soil it becomes heavier

pH TEST RESULTS

A few of the reasons for monitoring pH are as follows:
• Plant nutrient availability is pH dependent. pH 6.5 is generally ideal.
• Certain nitrogen fixing microbes won't live in pH 5.8.
• Early growth plants will respond to alkaline sprays, *i.e.*, pH 7 to 7.4.
• Fruit, root, or seed producing foliars require acid pH 6.4.
• Rain water in equilibrium with carbon dioxide will have a pH of 5.6.
• Acid rain has been recorded as low as pH 3.0.
• Some animal urine and saliva pH should be adjusted by diet to 6.4.
• Soil acidity is caused by free, active hydrogen ions. In very acid soils aluminum displaces reserve hydrogen on the soil particles, thus contributing to greater acidity. The Al^{+++} impact on acidity disappears when pH is 5.5.
• Drying a soil at a temperature above field conditions will increase soil acidity. During later parts of the season organic acids produced by microbes will be at a higher concentration.

In order to raise pH, use alkaline substances diluted in water — KOH, $Ca(OH)_2$, baking soda, NH_4OH, $CaCo_3$.

In order to lower pH, use acidifying substances diluted in water — vinegar (Acetic acid), citric acid, ascorbic acid, phosporic acid, sulfuric acid.

Substances which resist a change in pH are called buffers. Some soils are excellent buffers. Fine structure and organic matter improve buffering.

Buffering capacity may be determined by measuring pH of soil/water mixture before and after the addition of pre-defined proportions of: acid, alkali, and neutral salt solutions. By determining the difference between the various pH's and comparing with a chart showing how to interpret these readings will provide a greater understanding of meaning of pH measurements. — *Pike Lab Supplies, Incorporated* by volume. A bucket of pure sand gets very heavy. Add carbon and it "lightens up." Faltering density of soil nutrients means a smaller yield.

Let's get specific. About 2,800 pounds of soluble plant nutrients are needed for an actively growing crop in 5% organic matter soil. If the nutrient load is dropped to 1,500 pounds per acre, this shortfall becomes a limiting factor. Keep in mind the fact that carbon has an important role in holding nutrients in a given area. It also has the potential for increasing the nutrient density during the growing season by extracting nutrients from ionized air.

I have a base formula for small grains that works very successfully. Take one to three gallons of Thio-Sul and one and a half to three gallons of molasses, and flush these materials with water to get 20 gallons. The total formula reads as follows: ten to 15 gallons of 28% nitrogen, one tot three gallons of Thio-Sul, one to three gallons of molasses, 20 gallons with H_2O. Put this on just before you seed, the objective being to incorporate the mixture into the top inch or inch and a half of soil. This builds a magnetic field over the entire acre. It is much like a battery that is charged by the sun every day. At the end of the season a soil audit will reveal a nutrient level far higher than the one a similar audit will have measured at the beginning of the season. The Thio-Sul is 12-0-0-26, meaning 12% nitrogen and 26% sulfur. Here, as with most fertility instructions, deviation from the formula can usher in disaster. Too much nitrogen can paint the field green and push a lot of growth, but it won't support a small grain plant upright.

One day, it may be supposed, we will find more precise mathematical expressions for plant growth according to biological time. Pierre Lecomte du Nouy was able to do this with the process of healing of wounds. His equation enabled World War II physicians to follow the process of cicatrization and to calculate how long a surface wound would require to heal. His formula made it possible to compute the real physiological age of a patient and led him to a radical concept of biological time, a time quite different from the physical time of inert things. This time followed a different law, not chaos, a logarithmic instead of an arithmetic law. Time does not have the same value for a child as for a grownup.

Perhaps as we come to a more mature understanding of biological time for the corn plant, we will then be able to better measure the arithmetic of nutrition and growth as well as *real* yield.

9

THE BRIX INDEX

Photosynthesis is a term generally assigned to the business of explaining rather complex biochemical reactions covered by the formula

$$CO_2 + H_2O \xrightarrow[\text{GREEN PLANTS}]{\text{SUNLIGHT}} (CH_2O)+O_2 \quad \textit{carbohydrates}$$

yet neither word nor formula tells an entirely meaningful story. Nor do terms like *anabolism* and *catabolism*, which are breakdown definitions for metabolism. But there is a term that tells most of the "how it works" story concerning plants. That word is *sugar[s]*, singular and plural. During photosynthesis, sucrose takes its place in the leaves. Osmosis becomes operative, and cells with a high sugar content achieve turgor pressure. Early in his career, Carey Reams constructed a sugar index for plants that remained largely unknown until approximately 1982, when it was published in *Acres U.S.A.* The values presented drew nourishment from over 50 years of research. When this table achieved definitive form, Reams announced that he had discovered why the anatomy of insect control was seated in fertility management, and not in fabricating more powerful poisons. The answer to this revelation was securely locked in each brix reading, he said. Reams discovered that high brix readings on a refractometer meant a lower freezing point for plants and their fruit. The cycles decreed that citrus growers would be made to endure severe freezes in the 1960s, just as they will again in the 1990s. Two decades ago the ones to survive were generally clients of Carey Reams. Parenthetically it may be noted that soft rock phosphate is the one soil additive most likely to affect brix readings favorably. But the entire spectrum of anion-cation computed fertilization served equally as well. The reason was simple enough. Sugar readings of crops also meant mineral readings, the refractive reading being a good common denominator.

Here is the scale Reams worked out. A low reading in terms of brix means likely bacteria, insect and fungal attack. Average and good readings still suggest stress and trouble, but once an excellent brix reading has been achieved, it means that proper fertilization has conferred general immunity to bacterial, fungal and insect attack on the plant.

Serendipity often has a role in driving home some of nature's most absolute lessons. We were making homemade ice cream one day and we couldn't get it to freeze. I theorized that too much sugar in the mixture created the situation, just as high sugar in a plant prevents frost damage during a fringe night.

I have seen severe frost damage in alfalfa fields, and yet neighboring fields that had been foliar sprayed remained nice and green. This capacity for leap-frogging over a cold snap allows a farmer to lengthen his growing season.

That sugar content of an orange or a lemon or a watermelon can be measured by its shelf life is nothing but confirmation of brix values. A high brix orange will simply dehydrate, keeping a hard shell. One with a low brix value will decay.

The late Carey Reams liked to tell about a watermelon he entered at the local county fair three years in a row. I don't know

COMPARISON CHART FOR BRIX READINGS

PLANTS	POOR	AVERAGE	GOOD	EXCELLENT
Alfalfa	4	8	16	22
Apples	6	10	14	18
Asparagus	2	4	6	8
Avocados	4	6	8	10
Bananas	8	10	12	14
Beets	6	8	10	12
Bell Peppers	4	6	8	12
Blueberries	6	8	12	14
Broccoli	6	8	10	12
Cabbage	6	8	10	12
Carrots	4	6	12	18
Cantaloupe	8	12	14	16
Casaba	8	10	12	14
Cauliflower	4	6	8	10
Celery	4	6	10	12
Cherries	6	8	14	16
Coconut	8	10	12	14
Corn Stalks	4	8	14	20
Corn, young	6	10	18	24
Cow Peas	4	6	10	12
Cumquat	4	6	8	10
Endive	4	6	8	10
English Peas	8	10	12	14
Escarole	4	6	8	10
Field Peas	8	10	12	14
Grains	6	10	14	18
Grapes	8	12	16	20
Grapefruit	6	10	14	18
Green Beans	4	6	8	10
Honeydew	8	10	12	14
Hot Peppers	4	6	8	10
Kohlrabi	6	8	10	12
Lemons	4	6	8	12
Lettuce	4	6	8	10

COMPARISON CHART FOR BRIX READINGS

Limes	*4*	*6*	*8*	*12*
Mangos	4	6	10	14
Onions	4	6	8	10
Oranges	6	10	16	20
Papayas	6	10	18	22
Parsley	4	6	8	10
Peaches	6	10	14	18
Peanuts	4	6	8	10
Pears	6	10	12	14
Pineapple	12	14	20	22
Raisins	60	70	75	80
Raspberries	6	8	12	14
Rutabagas	4	6	10	12
Sorghum	6	10	22	30
Squash	6	8	12	14
Strawberries	6	10	14	16
Sweet Corn	6	10	18	24
Sweet Potato	6	8	10	14
Tomatoes	4	6	8	12
Turnips	4	6	8	10
Watermelon	8	12	14	16

how it tasted after three years, but he proved his point. The sugar content determines how well fruit will keep and its quality. Top quality produce will not rot. It will simply dehydrate. There is a saying that all generalizations are false, including this one! The tomato resists identification with the above general rule. Even an excellent tomato will resist dehydration. Still, a top quality tomato will have longer shelf life before it starts to deteriorate.

Green grapes have a pearly white transparent color. They should run about 26 to 28 brix on the refractometer, instead of 10 or 12, and be very sweet.

Viewed from almost any angle, the refractometer is modern agriculture's most valued tool. It enables the grower to perform an on-scene test that is often more valuable than the most sophisticated laboratory readout. Instructions for use are easy to follow. Two basic requirements must be met — a properly calibrated instrument, and a drop of juice from plant materials. The brix reading is simply the carbohydrate concentration in 100 pounds of juice stated as a percentage. Although the brix reading is loosely called a sugar index, it is really much more. For the higher the carbohydrate content in a plant, the higher the mineral content, the oil content and the protein quality. For instance, 100 pounds of a fruit with a brix of 20 translates into 20 pounds of crude carbohydrate if the fruits were juiced and dried to zero moisture. The 20 pounds — which also represents the brix reading — can be divided by two to factor out the actual sugar content, in this case ten pounds.

The brix reading gives a whole biography for the plant, if we develop the wit to read it. A faltering brix reading suggests low phosphate levels, and when the phosphate ratio to potassium is on target, brix readings will be uniform in the plant, top to bottom. Otherwise, the sugar will vary from the bottom to the top of a plant.

Properly focused, a refractometer has the capacity for bringing into view a sharp demarcation. The sharp line telegraphs the fact that the crop is low in calcium. But when the line becomes diffused, the acid is low, calcium high. We have already seen the shortfalls that attend low plant calcium. High plant calcium with sugar will deliver a sweet taste. This explains why a low brix, high calcium plant tastes sweeter than a low calcium plant, even though both have the same brix reading.

It takes art and science to mine the mother lode contained in the brix reading. Each clue takes the grower back to his fertility management, and to a clearer understanding of what it takes to grow field ripened crops that are relatively immune to bacterial, insect and fungal attack.

I have been checking a lot of crops with a refractometer and have been finding different values, depending upon the crop.

I checked some corn and soybeans in Nebraska and found some very interesting results. I found that crops which have been on a program for three years or more consistently had fairly good readings.

IRRIGATED	Regular	High-lysine
Leaves only	13	16
Leaves with rib	8	8
Stalk at ear height	8	11
Stalk near ground	5	7

HIGH-LYSINE, NON-IRRIGATED	
Leaves only	17
Leaves with rib	9
Ear milk	10

The above samples are from fields that have been on a program for several years and still have a ways to go. Ideally, the readings should be equal at all levels. In this case, the leaf readings were high enough to keep the corn borers out and prevent the need for spraying. Top corn producers in adjoining fields were aerial spraying once a week and spending $60 per acre for insect control that this account's field didn't need. I wonder what impact this spraying is having on the environment when it isn't necessary.

In order to keep all insects out, the stalk sap must be above 12 brix before it confers complete insect resistance. I have met a number of people who have spent a lot of time checking their

REFRACTOMETER DEMARCATION

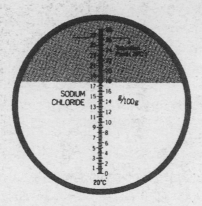

Properly focused, a refractometer has the capacity for bringing into view a sharp demarcation. The sharp line telegraphs the fact that the crop is low in calcium. But when the line becomes diffused, the acid is low, calcium high. We have already seen the shortfalls that attend low plant calcium. High plant calcium with sugar will deliver sweet taste. This explains why a low brix, high calcium plant tastes sweeter than a low calcium plant, even though both have the same brix reading.

crops with a refractometer and nearly all have found a direct correlation between lab leaf analysis phosphate readings and refractometer readings. The higher the phosphate readings the higher the sugar readings.

I recently checked high-lysine corn that had a lot of hog manure on its soil and found some very good readings.

Leaves only	17 brix
Leaves with rib	11 brix

Insect activity in this field was almost zero after you got in the first eight to ten rows of the cornfield. The neighbor's cornfields in the same area, where anhydrous ammonia and 18-46-0 had been used, had the following readings.

Field	1	2	3
Leaves only	9.5	9.0	8.5
Leaves with rib	8.0	7.0	6.0

These fields had numerous flying insects in them. When you shook the leaves, clouds of insects would rise into the air. I have seen this many times in the past ten years. I used to check the corn stalks at ear height and find readings varying from 3.5 brix to 8 to 9 brix, but never more than 12 brix. What puzzled me was that some fields had few insects and others had millions, yet the brix readings in the stalks at ear height were quite similar — and still the insect populations were quite different.

It wasn't until I started checking the leaf material only that I really began to understand the differences.

I have found that several cloudy days in a row will cause the brix readings in the leaves to drop from 17 to 18 brix in the leaf to below 12 brix in two to three days. The lower the humus content of a soil, the faster the brix readings will drop after several cloudy and rainy days.

Once the readings in the leaf drop below 12, insects move in and have a field day.

Excess nitrogen and potassium also can cause lower brix readings, because these two elements can enter the plant in non-phosphate form along with a lot of water causing the brix readings to drop. I wish I knew all the mechanics for what causes brix readings to rise and fall during a growing season, so that we could better manage insect problems.

Sugars are produced in the leaf during the day and translocated to the roots and back up to the various parts of the plant during the night. Many are converted to amino acids in this process, as well as to hormones and vitamins.

To understand a little more about sugar production in the leaf, let's look at the chloroplasts in the leaf which contain chlorophyll.

The chlorophyll in the leaf converts light energy into chemical energy. This is done by plant cells taking carbon dioxide from the air and water from the soil and converting the two compounds into simple sugars. The plant can do this only with the aid of chlorophyll and light. Chlorophyll uses light energy to form the sugar molecule.

To form the simple sugars, a molecule of chlorophyll located in a chloroplast absorbs light. The molecule becomes activated and loses an electron. *After it has lost an electron, it has a positive charge.* The chlorophyll molecule then can remove electrons from other substances — including water. The loss of electrons makes the water molecule unstable and it breaks down to release oxygen. When the oxygen is released, the hydrogen ion is free to combine with carbon dioxide from the air to form simple sugars.

Phosphate is the catalyst in the process of combining water and carbon dioxide into simple sugars. Potassium also plays an important role in this process by aiding in the closing and opening of the guard cells around the stomata.

It is also interesting to note that the chlorophyll molecule contains carbon, hydrogen, oxygen, nitrogen and magnesium. Magnesium controls the total amount of nitrogen in the leaf so an excess does not build up. I bring these points to your attention because of a recent foliar spray program used to increase the brix readings in raisin grapes. The following foliar spray program was outlined for a vineyard in California.

FOLIAR SPRAY
100 gallons/acre of water
1 gallon/acre of 0-20-0 (liquid phosphate)
1 quart RL-37

The first spray was applied two weeks pre-bloom. The second

spray was applied two weeks after the grape bunches had set. The third spray was applied two weeks before harvest.

The brix of the early fruit set was 4 to 6 brix and the harvest readings were 24 brix on the sprayed plot and 12 brix on the control plot. All of these readings were taken from the grapes and not from the leaves.

Petiole analysis indicates that during the rapid expansion of the berry from bulking to maturity grapes draw on phosphate quite heavily and don't seem to be able to keep up. This is why I think a foliar spray using a high quality phosphate source works so well along with a well prepared humate catalyst preparation.

Research at several universities has indicated that humates, when combined with phosphorus, can play a strong role in plant utilization and metabolism of phosphates. This must be true, because when you don't use the RL-37 with the phosphorus, the sugars don't rise the way they do when you *do* use RL-37.

The other key to the success of this spray program is the use of magnesium sulfate which speeds up metabolic processes and helps make sure there is enough magnesium for the chlorophyll molecule to keep the process of photosynthesis rolling to produce simple sugars. It is thought that humates also help liberate out of the soil carbon dioxide, calcium carbonates, and thus make carbon dioxide available to the plant through the roots for photosynthesis.

Humates are known to stimulate plant enzymes which further aid the production of simple sugars in the plant leaf.

I am not sure this formula will work in other plants, but it sure can't hurt to try. One of the limiting elements in this formula may be nitrogen, especially if the soil is deficient. If nitrogen is short, all you would need to do to this formula is add a little Bo-Peep ammonia to the mix, one quart per acre.

If you look at the formula for chlorophyll, you will notice nitrogen is needed to form the molecule. It is important to understand all these processes in order to build effective formulas to improve the brix readings of plants. It is always important to have a complete LaMotte soil test present when planning a foliar spray to see where the calcium readings are. If the calcium readings are below 2,000, it is often difficult to build a spray formula that will get lasting results.

I recently had an account call and tell me about his snap beans and their brix readings. He said he had leaf readings of 15 brix and pod readings of 9 brix. These readings were much higher than anything I have experienced before. He accomplished this by keeping the phosphates high in the soil by using a product called *Growzyme* and "fertigated" urea and calcium sulfate. He used 30 pounds of urea and 30 pounds calcium sulfate per acre three times after planting to get these brix levels. The best I have heard of before is about 10 brix in the leaf and 6 to 7 in the pod and stem.

I think the spider mite infestation in soybeans in 1989 would not have been a problem if we could have maintained leaf brix above 12 during the growing season. I have learned from another client that even though the brix of the leaves is above 12, plants can still be bothered by insects unless the calcium readings are kept above 2,000 pounds per acre.

I have been working with a tomato grower who used a refractometer during the summer, and this man's observations and refractometer readings confirmed what many others have told me: *when the brix readings of the leaves stay above 12, insects are not much of a problem.* The problem is that early season and mid-season when the plants are really producing seed or fruit, it is very difficult to maintain a brix reading of 12. This is especially true in soils that are low in basic fertility or soils that have serious problems with calcium:magnesium (7:1) and phosphate:potassium (2:1) ratios.

I don't know all of the ways that you can influence brix readings in plants, but I do know that they can be changed if you get the right nutrient to the plant at the right time. One of the best ways I know to get started toward better sugar values in crops is to have a fall program using compost, rock phosphate, non-acid treated, high calcium lime if needed, and ammonium sulfate.

For high-value truck farming crops, vineyards and orchards:
- one to two tons of good true compost
- 500 pounds high calcium lime (if indicated by soil test)
- 200 pounds ammonium sulfate

Then use liquid fish, seaweed powder and a little phosphoric acid in a foliar spray program every ten to 14 days during growing season. Here is a good spray formula:
- two quarts liquid fish 50% solids
- four ounces seaweed powder
- one-half to one pint phosphoric acid (85% food grade)
- 20 gallons of water

I have had many farmers across this country use the formula very effectively with little or no problems. The formula does not give fast results, but it improves the crop as the season progresses. A formula similar to this helped a farmer in Pennsylvania produce a 60-acre tomato crop that yielded more than 30 tons per acre, and helped a pumpkin farmer harvest 25 tons of pumpkins per acre at seven cents per pound.

The refractometer merely confirms an opinion I have long had, that muriate of potash should be eliminated from the agricultural scene. That one fertilizer has done more harm to the biological systems in our soils than any other product ever put on the market. Muriated potash was used during WWII to produce gun powder. When the war was over the industrialists found themselves with big factories and no market. So they dreamed up the idea of making agricultural land the dumping ground for their hydrocarbon trash. Initially these materials showed good response, always at the cost of steady soil deterioration and inevitable erosion via wind and water.

Anhydrous, muriate of potash, other types of salt fertilizers, all have conspired to create a false impression of what it takes to produce quality bins and bushels. At first the nitrogen input was 50 to 100 pounds, then it went to 200 pounds per acre — next 400 pounds, 600 pounds, finally 1,000 to 1,200 pounds per acre — D-Day. That's the day everything comes to a screeching halt. How long it takes to get to this soil crash depends on the organic matter or carbon in the soil.

I recall an Indiana farm — this man had a $150,000 brick home and a machine shed with almost factory capability. In the mid-1970s, he reached 200 bushels of corn per acre by pouring on more anhydrous, more muriate of potash, and more 18-46-0. By

THE PHOSPHORUS-BRIX CONNECTION

I have just received some more data and refractometer readings on grapes that are fascinating. This is information from a student in California who ran leaf tissue analyses against the refractometer readings.

Leaf tissue analyses ran from .1% to 1% phosphorus during the growing season. Once the leaf tissue analyses ran below .35% phosphorus, the brix reading would not stay above 12 brix in the leaf. If the leaf tissue analyses stayed up in the .4% to .6% range, the refractometer readings would stay around 13 to 14 brix during the filling stage of the grapes.

He also observed that there was a close tie with the leaf tissue magnesium reading and refractometer readings. He found that the magnesium readings needed to stay above .3% and better yet between .4% and .5% He found that if he had the higher readings on the magnesium, he could have lower readings on the phosphorus and still maintain readings at 12 brix or better as long as the phosphorus readings did not drop below .33%.

the late 1970s, he couldn't get 80 bushels per acre, and it wasn't because of drought. The refractometer could have told this farmer what was happening before it happened. I helped this farmer change his program, and before long he was back to 180 to 190 bushels, and his soil was recovering. Without help, that farm was strictly a sunset operation.

The refractometer — in its subtle way — warns against the use of dolomite limestone. When purchasing limestone, no product should be used on the soil if it contains more than 5% magnesium. If the source says the information isn't available, then request a laboratory analysis or look elsewhere. I have in mind a bunch of Michigan farms on which the wrong limestone was spread at the rate of two tons per acre. Years later, those fields still won't grow a crop. That's why it is mandatory to check the analysis.

My best recommendations on the use of anhydrous ammonia is — don't! If it is used, never apply more than 40 to 50 pounds per acre at one time. Some growers are using aqua ammonia and adding molasses. This works quite well, and much less is needed. I have discerned a basic rule when it comes to the use of anhydrous. If you use 28% liquid nitrogen properly, the successful corn corp that normally takes 180 pounds in an area can be grown with only 110 pounds. If a better job is accomplished, the range of 80 pounds per acre should suffice. During the most recent season, I grew a 175 bushel corn crop in my area with only 40 pounds.

During the season, any day of the week, the brix reading can give all the late breaking news. More important, it poses questions and often stays on for answers.

The use of elemental sulfur is fraught with danger, even when an overload of magnesium suggests its use. Sulfur, to be effective, needs to be in a sulfate form. Sulfur has to be worked on in the soil by bacteria in the presence of water. Only then can it be used by biological life.

The lessons go on in the rest of this book. But this interlude on

THE REFRACTOMETER

Hand refractometers are handy measuring instruments with which anyone can measure the concentration of an aqueous solution. Plant juices offer a concentration of dissolved solids, sugars, and amino acids. Brix readings can define the probability of yield limitations, and high brix readings in effect confer immunity to fungus, bacteria and insect attack.

the refractometer has meaning both ways, looking back to everything covered so far, and forward to the end of the message. It is ironic that no university seems interested in this little tool that could surely revolutionize agriculture. Perhaps a dozen have been asked to research the proposition Carey Reams stated so clearly. Even after the brix indexes were published in *Acres U.S.A.* — and consultants made it a part of their manuals from Europe to Australia — an academic reluctance has held down the concept with a lead glove.

10

THE CALCIUM CONNECTION

"We use limestone to supply calcium, rather than to remove soil acidity." So wrote the late William A. Albrecht when he was chairman of the Soil Department, University of Missouri. And in so many words, he made the calcium connection. Volume III of *The Albrecht Papers* is the soundest vintage appraisal of calcium. *Fertilizers and Soil Amendments* is another text anyone in farming should have for its valuable information on calcium. Albrecht noted that this belligerency toward soil acidity has carried along calcium as a fertilizer material, but "Ash analysis of plants leaves confusion about the services of calcium."

Our own Milhouse Unit evaluations of one atom of calcium is 10,750, making it a middle of the road energy source. This computation is based on calcium as an average in terms of Milhouse Units. The lowest possible energy value for a single atom could be 540. That is why the energy level of calcium is low in some soils even though there are more pounds per acre compared to a second soil with fewer pounds per acre.

The atomic weight of calcium is 40. This means a calcium atom will have 40 rings around it. Since calcium is anionic, it has a negative charge on its electron shell. Simply stated, the anionic value of the electron shells could be at 40. The lowest value of a positively charged substance is 500. The plus charge of 500 added to the negative charge of 40 yields a total of 540. If calcium that has been depleted of its energy field is applied, obviously it will take more of it to do the same job as calcium still possessed of its energy flush.

Calcium must be understood from an energy point of view, and this must not be confused with wet lab chemistry. This often pushes the Biological Theory of Ionization at loggerheads with conventional laboratory reports. Calcium in its purest state has a pH of 14 and is considered a non-conductor of electrical current. The neutrons which rotate around the proton in calcium move and still construct the calcium element. As I pointed out earlier in *The Farmer Wants to Know*,

> The number 7 is considered to be neutral because water has a pH of 7. Water is a solvent, and it is the best soil plant food catalyst known. Soil elements or compounds whose electrons rotate faster than those in water are now classified as an acid in soil nutrients. Those elements of compounds whose electrons rotate slower than those in pure water are said to be alkali. In the purest scientific sense, this is a contradiction, but I am now stating what is considered acid or alkali regardless of the true scientific classification. Consequently, a false impression results relative to what constitutes sweet and sour or acid and alkali soils. The speed with which the electrons actually rotate in the soil compound substances has very little to do with direct plant feeding and the pH reading. The conductivity of the plant food proteins does have much to do with the rate of growth in plants. The pH reading does greatly affect the forming of plant food proteins which are direct plant food. In the molecular structure the proteins become more or less good or poor conductors of the electrical current passing over the crust of the earth. THE pH READING IS AN ELECTRICAL MEASUREMENT OF THE RESISTANCE BETWEEN THE ACIDS AND ALKALIS PLUS THE MAGNETISM OF THE ELECTRICAL CURRENT FLOWING OVER THE CRUST OF THE EARTH.

The rate of this flow is largely determined by the molecular structure of the soil. The difference in elements and compounds is only the difference in the number of anions or cations in orbit in the molecular structure. ELEMENTS ARE TRULY ALKALIS WHEN THE PROTON IS THE NUCLEUS AND THE NEUTRON IS THE OUTER ELECTRON. WHEN THE NEUTRON IS IN THE CENTER AND THE PROTON IS IN THE ELECTRON THEN THE MOLECULAR STRUCTURE IS AN ACID OR METAL WHICH WILL CONDUCT ELECTRICAL CURRENTS.

When cations are the center of the atoms and the anions are the electrons, then the substance becomes a poor conductor of electrical currents. Opposite factors have greater attraction for each other when the space is greater between the orbiting electrons. For this reason, plants grow better in warm weather than they do in cold weather. Naturally there are extremes in heat and cold. When the temperature rises to about 90 F. the speed of the electrons becomes so great that plant roots cannot magnetically attract the plant food elements and hold them because of their momentum. The greater the speed, the greater the heat becomes and the greater the soil's holding power of heat. This is true of both high and low pH readings as long as this exists.

The total amount of calcium or potash found in soils by flame photometer soil test methods means very little. This is because the total amount of elements present in the soil has very little to do with the available amounts of plant food. Energy can be accurately measured only by liquid soil test methods. The cation nutrient and anion nutrient molecular count can be tabulated to accurately predict the timing of the release of the plant food energy. This can be done so that more fertilizer can be added before all the plant food energy is used and any damage is done. *Direct plant food energy cannot be accurately calculated by flame photometer soil test analysis.* Then, you might ask, why is this soil test method used? The answer is that it is a quick and very cheap method of making soil tests. This kind of test is better than nothing at all, but it falls far short of what other test methods could do for persons involved.

Since calcium in the electron shell carries a negative charge for the purpose of energy computation, each of the electron shells contains a potential of from one to 499 Milhouse Units. To compute the highest value, we must multiply 40 times 499. To get the highest value for the nucleus — which has a positive charge — the arithmetic takes on a new multiplier, one usually based on the range of 500 to 999. Taking the highest number, the two computations added together yield a sum of 20,959. If calcium is added at the lowest

computed value based on pounds, the equation would be
$$20,959 \div 540 = 38 \text{ pounds.}$$
This means that one pound of calcium at the highest value would equal the energy of 38 pounds of calcium at the lowest possible value.

This may sound confusing, but field experience validates this dead reckoning. I have seen case reports in which the highest yield of field corn had the lowest calcium reading. The only explanation is that the calcium was high in energy.

It would be impossible to have all the calcium at the maximum energy level. Nature has decreed that it must be synchronized at some level in between. The point here is simply that calcium measured in pounds and tons is often meaningless information.

There are approximately 5,000,000 atoms in one drop. If we calculate the highest energy value for one drop of Ca- and the lowest energy value for one drop of Ca-, 104,795,000,000 would be the value for the former, 2,700,000,000 would be the value for the latter. If all the calcium was at the lowest level, it would be unable to give up any energy whatsoever. Unlike many farmers who consume their capital in order to stay in business, calcium cannot consume and destroy itself. But when calcium has been stripped of its energy, it can no longer service crop requirements.

Calcium recovers its energy when the weather switches the soil back and forth. As the soil moves into an anionic charge, this supports a buildup of anionic soil elements. When the soil moves the other way, it releases energy, and in this way the system is kept mobile.

What our instruments and figures are saying, in effect, is that the farmer no longer needs to operate either by the seat of his pants, or according to rules and soil audits that are faulted. Our foundation concepts explain why one fertilizer will fit a specific field situation better than another. Our readouts and numbers tell us that when we go to water solubles we have immediately all the energy "in solution." The solubles are necessary in order to get the energy to release in a timely way. That is why some numbers declare the ability to charge calcium anions temporarily, then pull the energy back. By working energy back and forth the net result is an increase in the total units of energy on that acre of land. As this increase takes place, there should be more flow than can be given up for the crop.

The question is often asked, *Is the sun charged with cations and anions?* We have explained how an anionic charge comes in from the sun at an angle, crashing into the Van Allen Belt, giving the planet its spin.

Parenthetically, it must be noted that there are eccentric and even tilted planetary orbits. Uranus and Venus rotate in the opposite direction of the rest. We don't know exactly why. Some satellites of planets are also in retrograde motion, and there is even distribution of angular momentum among the satellites. The moon has a lower density than the earth, and the heaviest elements are predominantly in the smaller planets.

Cations and anions from the sun are a fact of life. Both depend on weather, temperature, moisture and wind. The anions achieve a 499 energy level, and then they become cations. As the anion becomes a cation, it switches the rotation of the electron shell.

Calcium, added to the soil, can take energy away from the crop while it is blending with the soil complex. When this happens, it is important to add another source of cationic energy for crop use. A carbon molecule attached to the calcium does not permit as much energy escape because there is a lot of energy stored in carbon. There is also more energy in 100% water soluble calcium than there is in talcum powder fine lime, and it is more available. It does not last as long, for which reason business economic considerations enter the picture. The best available water soluble calcium is calcium nitrate.

When doing the pencil work on a fertility program, I always look for a calcium source, or something to activate calcium. William A. Albrecht once called calcium "the prince of nutrients." It was probably the only error of his career. He should have said, "the king of nutrients," and high priest, or whatever. It always has been and always will be that primary nutrients are worked for energy release–basically, working a positive against a negative. In the soil there has to be a positive and a negative, or no current

A DESIGNER

One cannot be exposed to the law and order of the universe without concluding that there must be design and purpose behind it all ... The better we understand the intricacies of the universe and all it harbors, the more reason we have found to marvel at the inherent design upon which it is based . . . To be forced to believe only one conclusion — that everything in the universe happened by chance — would violate the very objectivity of science itself ... What random process could produce the brains of a man or the system of the human eye? . . . They [evolutionists] challenge science to prove the existence of God. But must we really light a candle to see the sun? . . . They say they cannot visualize a Designer. Well, can a physicist visualize an electron? . . . What strange rationale makes some physicists accept the inconceivable electron as real while refusing to accept the reality of a Designer on the ground that they cannot conceive Him? . . . It is in scientific honesty that I endorse the presentation of alternative theories for the origin of the universe, life and man in the science classroom. It would be an error to overlook the possibility that the universe was planned rather than happening by chance. — *Wernher von Braun* flows. The primary negative is calcium. Almost all other plant nutrients are positive.

Heraclitus, the Greek philosopher who said "all is change," must have had agriculture in mind. Whatever a farmer does, two weeks later there can be a *change*. Success for the entire season depends on being aware of change. "Two weeks later" might as well be written across the farmstead gate in bold letters. Two weeks later the energy availability of calcium may be down the tube, with serious trouble to follow. A heavy rain can take out those solubles, and it will if there is not enough organic matter with its carbon to hold them. The dry fertilizer material may hold a little longer, but if biological activity is absent, it won't break down at all. In order to jack up activity, I use a little calcium nitrate once in a while.

A smattering of information on calcium sources won't do. For this reason, it becomes necessary to detail what the market has to offer, together with some commentary on the validity of each type. Calcium nitrate has been mentioned, 100% water soluble. It is not uncommon to use 100 to 200 pounds per acre on alfalfa, spread on a dry broadcast basis. The same product in a water solution can be sprayed on at the rate of four pounds of actual calcium to the acre with 20 gallons of water and do the same thing. A repeat application with every cutting is usually indicated. The point here is that the 200 pound application represents a complete loss of energy if there is a heavy rain. In such a situation, a double anion would develop, and anions go up. Also, if you load that much energy onto the soil, there is too much heat and too much energy release at the same time, both representing an eco-nomic loss. If more than four pounds in 20 gallons of water are sprayed on the alfalfa after cutting, the crop will suffer burning.

Like attracts like. This is why gold lies in veins, as do concentrations of silver, iron ore and calcium in lime pits. When a small amount of calcium in 100% soluble form is introduced into the soil, its energy power is far greater than any of the elements surrounding it. If there is calcium in the soil and some moisture, the new input has the tendency to draw the insoluble calcium out of the soil and into the soluble. I have done this and I have verified this phenomenon with LaMotte tests. Ten days later the same test result on the same field — where I sprayed on the calcium — would be 10,000.

This may be something like eating too much Thanksgiving dinner. You can stuff yourself — much like putting on too much soluble calcium — and be miserable for several days until you recover. I think we do much the same thing to crops.

In Moses Lake, Washington, one of my friends had a test plot with peas for the canning trade. This was set up with a 160-acre pivot set up to cover only four acres. They covered one acre with a tarp. Each time they sprayed four pounds of calcium nitrate, keeping a weekly schedule. One of these areas got four treatments, one got one treatment, and one got three with four pounds being used each time. Every time there was a four pound application there was a 500 pound increase in peas harvested. One area had 2,000 more pounds of peas than the other.

Based on this information from Bill Johnson, Soil Spray 8, Moses Lake, Washington, I started making a four pound application in a water carrier on alfalfa fields with similar results.

It is possible to mix calcium nitrate with a lot of products — Thio-Sul and 28% nitrogen included. There is one caveat. If you have a very inferior grade of phosphorus, your mixture will turn into the thickest bowl of jelly possible. If you ever make the mistake of filling a sprayer tank with this heavy syrup, you may have to abandon the equipment.

A second warning is also in order. If the crop has big leaves — such as the pumpkin plant — four pounds may be too much. Those leaves are extremely sensitive. Much the same is true of many deciduous trees. Grapes also are extremely sensitive. Leaf application on peas and alfalfa works well, however.

I have never used calcium nitrate on corn leaves. I have always applied it between the rows near the roots.

On alfalfa, there is no such a thing as "too soon." The sooner calcium can be returned to the scene, the faster the regrowth. Some farmers have a spray bar on their cutters and spray the solution even as they cut. There is such a thing as too late. If the alfalfa gets three or four inches in height, the leaves and edges will start to exhibit burn. This means more water should be used, or less calcium nitrate should be used.

There is a human problem with many of the instructions for proper use of fertilizer materials on the market. Too many farmers like to substitute their judgment for the judgment of the consultant. Too many think that if a little is good, more will be better. Since directions have been usually worked out over years and with special knowledge in tow, the point must be made that any change in directions by the farmer absolves the consultant and the fertilizer fabricator of any responsibility.

The next calcium on our roster is calcium carbonate — generally known as ag lime. In this compound the carbonate and the oxide are bonded together. Spread on an acre of soil, calcium carbonate usually is applied at between 500 pounds and two or three tons per acre. Sometimes dry blends use 100 to 150 or 200 pounds per acre very effectively. A warning is in order — again! Always get a sample from the quarry, and be certain the delivered product is the same as the sample. Some lime materials are toxic.

Gypsum — calcium sulfate — is another calcium source. No more than 500 pounds per acre should ever be applied in any one year. It is an excellent amendment when the soil has a high salt content, surplus magnesium, and in general is tight and compacted. Irrigated areas, especially in the hot, dry areas of California and Arizona, tend to build up salt. In some of these cases calcium sulfate has proved itself in pulverizing and loosening the soil. I know of applications of up to two tons per acre. This procedure cannot be recommended in higher rainfall areas. Calcium sulfate can be blended with ammonium sulfate, but the blend must be spread quickly. Ammonia and sulfate are both cationic. Calcium is anionic. Allowed to do their chemical thing, the elements in this mixture will set up like concrete if not spread promptly, certainly not a day or two after blending.

For all practical purposes, over 80 to 90% of all nutrient uptake is achieved in the top three or four inches of soil. There are exceptions, and to iron them out we like to think in terms of the top six inches of soil. Calcium hydroxide has extremely special uses, which I can best explain by giving a case report that involved Carey Reams. He was a consultant for a farm with 100 acres of cabbage. The weather forecast promised a heavy freeze, almost guaranteed to wipe out the crop. Reams knew that a heavy freeze would switch the soil to strong cationic in the absence of an anionic situation. Calcium hydroxide is a very powerful anionic substance.

Reams had the farm workers trickle calcium hydroxide along the side of the cabbage rows in a line on top of the soil. These cabbage plants wilted and fell over, for all practical purposes they appeared to be dead. But within 24 to 48 hours they came back, and by the end of the week they were growing normally. Reams always related how some workers refused to go back to work because the dead were supposed to stay dead, and Reams was raising the cabbages from the dead.

Actually, Reams merely used basic chemistry. Early in the spring he put on 200 pounds of calcium hydroxide per acre to gain two or three weeks on the growing season. Reams harnessed the same device to the cold snap. Farmers who achieve a certain sophistication in fertility management will likely find special uses for calcium hydroxide in the seasons to come.

Finally, the product Aragonite rates some attention. It is merely pulverized seashells from off the coast of the Bahamas. It has a tendency to be soluble and very pure. It can be used much like ag lime — calcium carbonate or calcium hydroxide. Aragonite is the trade name.

The bottom line to any discussion is always a codicil to what Albrecht, Reams and other astute investigators have pointed out for half a century. A change in the cation-anion balance of an atomic structure means a changed pH reading. And for this reason pH does not give a valid answer to the question of when calcium is needed for crop service.

11

LET THERE BE LIGHT

According to *Genesis* 1:3, God said, *Let there be light*. And there was light, moving at a speed in excess of 186,000 miles per second. Although it took man centuries beyond comprehension to discover some of the meaning of that pregnant sentence, we now realize that every atom in existence is ruled by *Let there be light*, that undulatory, vibrational creation, with its fixed speed of 186,280 miles per second, to be exact.

It now seems strange that we have to go back to the *Bible*, the best selling book of all time, to fully understand the Carnot-Clausius Law of conservation of energy, that enigmatic statement to the effect that energy can neither be created nor destroyed. The anions and cations give precise properties to all the elements, including those we know figure in crop production — and all have properties related to the speed of light. All have motion at speeds that range from well below the speed of light, some in equilibrium with light 186,280 miles per second, some faster than the speed of light. Each atom has its position, its energy levels and its computations for field application. This proposition requires iteration and reiteration as we continue to examine fertility products suitable for crop maintenance at optimum levels. From that opening sentence on light to the last verses of the *New Testament,* we see constant reminders that life — plant, animal and human — is meant to be lived in health, and that "mischief [Job 5:6] comes not out of the earth, nor does trouble spring out of the ground, but man himself begets mischief as sparks fly upward." Indeed, anions fly upward because of mistakes in agronomy. Rock phosphate, a compound, represents a miracle in bonding together phosphorus and oxygen. Soft rock phosphate holds locked in its molecules a great range of essential plant minerals. Rock phosphate also contains calcium, albeit not in significant amounts.

Materials useful to making the proper anion-cation connection turn up in some unlikely places. Basic slag is a byproduct of the iron ore smelting industry. They use calcium in the smelting kettles to keep the molten metal from spitting out the top. In the process the lime picks up iron and trace metals. The recommended application rate is 500 pounds per year, which will put about 25 to 40 pounds of actual iron into the soil if that is needed.

Tricalcium phosphate is a highly insoluble calcium. I would not even consider using it unless the price was extremely right and I had a biologically active soil. Then there is bonemeal. Bonemeal is used in a lot of gardens, it is not economically practical on any other scale.

There are a number of new products on the market, many of them calcium based. There's one called liquid calcium. It is made with calcium nitrate — as most of the special products are — with a chelating agent added. In fact, there is a corn starch chelating agent that some fabricators mix with it. The basic source for all the liquid calciums that are sold in gallon jugs is calcium nitrate. And in certain situations they work quite well. The basic one on the market is blended with a sugar-based product, calcium nitrate and urea. Calcium oxide and calcium carbonate also go together quite well. Generally speaking, lime from the pits means ag lime.

As suggested earlier, soft rock phosphate is a totally unique product. It takes a little know-how to use it properly in various parts of the country. If there is a single product that seems to improve crop quality, it is soft rock phosphate — if managed properly. There is a problem with soft rock phosphate most farmers cannot easily resolve. With soybean and corn prices determined as they are, soft rock phosphate often is not economically feasible. The product also has technical problems in spreading. If it can be spread evenly over all 43,560 square feet in an acre of soil, it will form a gelatinous glue over the surface of the land. One pound of this material can cover theoretically 7.5 square miles as a single sheet. If adequate coverage on an incline ever gets wet and there is additional field work to be accomplished, I doubt that it would be possible to mobilize enough traction to cover the incline. I had 500 pounds applied to a Colorado farm. When the farm operator ran irrigation water across the field, the water was at the far end three to four hours sooner on the treated acres than on the next field.

I have counseled a family with the largest family-owned vegetable farm in the United States, and I have spent some time on that farm. At one point they put Calphos on land under pivot irrigation. Instead of irrigating the land every day in order to get spinach to market, they watered it every third day and harvested a bigger yield. In this case, one ton per acre was applied. However, it must be remembered that spinach is a very expensive crop with a high per acre return. At the time, soft rock phosphate was delivered at approximately one-fourth the price it is today. Calphos prevents leaching in the soil. It battles the two engines of erosion, wind and water, and constructs a fair measure of drought resistance. I used it shortly before this writing on a 320 acre field in South Dakota at the rate of 250 pounds, and then I added gypsum, or your calcium sulfate, and ammonium sulfate with a slurry spreader. I observed an interesting thing in the process.

Monsanto Chemical Company hires college kids during the

summer for the purpose of monitoring insects. When insect populations reach a predetermined level, the word goes out to Extension agents who in turn put out bulletins for farmers to do their insurance spraying. This is how the apparatus of government is turned into tax-paid salesmen for the makers of toxic genetic chemicals.

One of these young investigators crossed over into this cornfield without permission. As a consequence he was asked promptly what he was doing. He said, *I'm out here getting a bug count for Monsanto,* and your field is on my block. It had rained, and the corn borers were all dead. He asked why they were dead, and then he looked down and saw a little water—*Oh well,* he said, *they drowned.* Three or four weeks later a new hatch was out and the insects were all dead again. The reason is that when you have insects in a field that has high energy and a high sugar content in the crop, alcohol is produced. A human being can consume alcohol with moderation. An excess can cause diarrhea, but diarrhea in a human being is nothing compared to the same malaise in an insect. In the last case it is a major catastrophe because the insect dehydrates rapidly. If that doesn't get the insect, it becomes inebriated and falls to the ground where active bacteria promptly have an insect lunch.

Prairie soils do not often call for rock phosphate. If they have a good nutrient level and clay soil base, use very little or none at all.

In Florida it takes about a ton of Calphos to the acre to build a magnetic field because most of the soil has pure white sand as a base. Further north, depending on soil conditions, less and less is indicated. In south central Minnesota, analysis will order up between 100 and 250 pounds. It will be very effective at between 100 to 150 pounds per acre — as a part of a blend, of course, because it is almost impossible to spread at the rate suggested above. The next to the bottom line is that it will improve the magnetic field considerably. The bottom line is the lack of economic feasibility.

If the crop has a higher market value — such as alfalfa — Calphos might be indicated if facilities are available to blend ammonium with a little lime, either ag lime or gypsum. I cannot stress too strongly the need to check each time whether it would be advisable to add humates to the blend.

If the soil has a high organic matter complex, the humate will kick out, and no appreciable effect will show up. Adding the humate material to a slurry seems to be far more effective then spreading it dry. In Idaho, where I worked with it some few years ago, we applied it in a slurry in the fall. The slurry spreader broke down. As a consequence, the material froze up, and the farm operator never got back to spreading. In the area he managed to cover, whatever crop he is growing will be a foot taller and the yield unbelievable to this day. The area to which the humate material was applied dry has never exhibited a similar response.

When you take a fertilizer product and put it into a spin, it will have one effect. If the material is put into an opposite spin, it will have a different effect. If I compound a product at my laboratory and put it in a vitamix, and spin it one way — and run a standard LaMotte analysis — nitrate nitrogen shows up superbly. Sometimes there is even a small phosphate reading. When I spin the same material in a vitamix in reverse, there will be no nitrate nitrogen reading, and a reading for ammonial nitrogen instead.

I have pursued some of these experiments to their appropriate conclusions. In Idaho I determined that the spin exacerbated success or failure. I have mixed stack dust lime and some few other ingredients and gave the mixture a spin one way, with no resultant smell. A reverse spin resulted in an ammonia odor that could hardly be endured.

When you agitate your liquid mixes, a kicker motor often becomes an agitating motor. I have this vision that in certain situations — if there was some way to reverse the motor — we might see things happen we never dreamed possible.

Nitrogen is an isotopic element. When the spin is one way, it is negative; another way, it is positive. And it shows up that way with lab analysis.

There is available a hard rock phosphate that in most soils would not be effective. But in hyperactive biological soil it delivers an excellent response. It breaks down slowly. Normally this phosphate form is mined then treated with acid to make the standard product more soluble — your 18-46-0, for instance. There is also an 11-52-0 and there is super phosphate — 0-20-0. The 0-20-0 product is made by taking hard rock phosphate and treating it with sulfuric acid. Sometimes fabricators treat the rock with phosphoric acid after they make 0-20-0 to get 18-46-0. This avoids the high cost of shipment. Triple superphosphate — or 0-46-0—rarely rates recommendation from my consultation chair. That product can work in a blend, however, if the mixture has a little lime, a little ammonium sulfate and a humate involved, or is used along with a compost dressing. It will account for a good energy release when cationic energy is required.

There is a chemistry involved whenever anything is put into the soil, inorganic, organic, salt form, whatever. Rock phosphate is called tricalcium phosphate, and this means it has three calciums, or three negative charges for bonding. This makes it more difficult to detach from fixation than would be the case with dicalcium phosphate, which has only two charges — thus *tri, di*! Last, there is the water soluble monocalcium phosphate, which means that as a consequence of acid treatment this form has only one remaining bond.

It may be that a brief quotation from *An Acres U.S.A. Primer* is in order at this time.

If you're over 6.5 pH, and you want to farm organically, for good and obvious reasons, you're in trouble. You probably shouldn't be using non-water soluble phosphorus because the soil does not have enough acid to free it up. If a soil system has a pH of 7.5 the farmer probably shouldn't be using the *di* forms. He should go to strictly *mono* forms of phosphate. Any farmer who doesn't take into consideration the importance of the active hydrogen ion as being the most important thing to work with thereby authors his own failure.

Acid treatment merely means rock phosphate is being converted from tricalcium phosphate to monocalcium phosphate, and that this highly unstable form is subject to natural reversion back to the stable tricalcium form. The rate of reversion differs. The pH, the free calcium in the soil, the organic matter — all figure in this rate of reversion. But it is safe to say that 75% of the monocalcium phosphate reverts back to stable tricalcium phosphate within 90 days. In some soils the reversion takes place

IONIC CONDUCTIVITY

Ideal is 150 to 190 uS or microSiemens (formerly micromhos/cm).

ERGS as defined by Reams is Energy Released/Gram of Soil.

Equal volumes of soil/DI water are gently mixed to dissolve those most readily available plant nutrients. An AC conductivity meter is used to measure the quantity of electrical charge transferred at either probe due to quantities of individual ionic species which have sufficient mobility to reach the oppositely charged probe during the half-cycle frequency of the meter. A large cluster of atoms with a small ionic charge will have too much inertia to move to a probe contact or to be readily available to a plant rootlet is the basis for this ERGS test.

Ionic concentration is dependent on health of soil, plant, and state of materials added in past.

Perfect water has an equilibrium dissociation of H_2O into H+ and OH- ions which produces an ultimate lower limit of conductivity of .056 microSiemens.

Reasonably good quality distilled water will have a conductivity of 1 to 5 uS. Good natural spring water will have a condition of 15 to 25 uS. Good well water will be less than 100 uS. Water of 1,000 uS is not drinkable on a routine basis! Water distillers are available to purify the poor quality water.

ERGS of pure sand and water will be less than 10 uS. The ERGS of a good natural woods earth soil will be 100 to 200 uS. If a soil has its nutrients tied up or complexed, then the ERGS will be low 2 uS and plant growth reduced. Some crops such as corn may be pushed to greater yields by bringing the ERGS to 400 uS. A baseline reading of ERGS may be established by gathering a soil sample in

the early spring after the fall and spring rains before the bio-life has been activated with rising temperatures. Salt residues and unutilized plant nutrients results in baseline ERGS of 25 to 600 uS. If soil ERGS equals 1200 uS most plants won't survive. 1:1 v/v H$_2$O/Soil sample may be used for: pH, ERGS, pNa, rH or ORP (redox potential).
— *Pike Lab Supplies, Inc., Strong, Maine.*

within hours. As soil conditions worsen, release of nutrients from rock phosphate worsens, and the chemical amateur becomes married to buying salt fertilizers, each go-round worsening still further the structure of that soil.

The water soluble phosphates are simply water soluble, not acid. But they are a poor substitute for having the proper pH with calcium, potassium, magnesium and sodium in equilibrium, and from an economic point of view they take on ripoff dimensions.

First, the soluble phosphates come from rock phosphate in any case. By treating, say, 1,400 pounds of rock phosphate with 1,200 pounds of sulfuric acid, the fertilizer industry gets 20% superphosphate — the tricalcium phosphate form being converted into water soluble monocalcium form. This chemical reaction causes 20% superphosphate to be represented by about 45% monocalcium phosphate, and 55% calcium sulfate, or gypsum. This means the bag of 0-20-0 contains about 45 pounds of water soluble monocalcium phosphate, which is presumably desired, and about 55 pounds of calcium sulfate, which may or may not be desired, but which farmers are frequently not aware of.

The fertilizer rating 0-45-0 is quite different material. Farmers who see symptoms of phosphorus deficiency sometimes think a higher rating is the answer, and this one comes styled triple superphosphate. Here the acid used to do the etching is phosphoric. This eliminates the calcium sulfate in the bag, calcium frequently needed to kick up the calcium reserve, sulfate needed to complex an excess of magnesium. By invoking the hot-dog concept of plant nutrition a much needed nutrient might be eliminated exactly when it is needed.

Ammonium phosphates such as 8-32-0, 11-48-0, and so on, also involve concentrated phosphoric acid in the processing, and this provides a handy outlet for otherwise unsalable fossil fuel company byproducts.

We have mentioned the *di* forms of phosphate, diammonium with a P_2O_5 rating of 48 to 53. What do the microbiologists say about non-straight forms of nitrogen compounded with phosphorus as a mixed fertilizer? Apparently the livestock in the soil work fairly well with this mixed fertilizer form *di* forms are synthetics, of course, and the question surfaces as to whether synthetics have any place in a good biological farming system. To pose that question is to suggest that there is an answer. There is, and it has to be answered when crops are born. If insect, bacterial and fungal attacks threaten a crop, a mistake has been made. Such a mistake is equally damning in its finality whether a soil system is managed according to the precepts of eco-agriculture, or in compliance with the chemical gambit. Suffice it to say that mistakes in terms of biology are inherent in straight chemical agriculture, hence the highly touted lines of toxic weed control and rescue chemistry.

This passage has merit and puts in perspective my own anion-cation evaluations. For now it is enough to note that diammonium phosphates best go on at the rate of 100 to 200 pounds per acre. Triple super, when indicated, will go on at anywhere from 100 to 300 pounds per acre, and 0-20-0–a very hot product because of the sulfuric acid treatment — rates a 200 pound per acre application as a maximum at any one time.

Liquid phosphoric acid — when it is good quality stuff–will run 75 and 85%. It was a byproduct now generally dried up because of pressure on the big soap companies to get the phosphate out. *Farm Chemicals Handbook* lists in some nine pages of fine print the names of fertilizer firms and their products, almost all of which can be used in a sound ecological system if the anion-cation properties are understood and measured before use. Those same products spell insect, fungal and bacterial disaster when used as most farmers use them, by the seat of their pants.

The point here is that fertilizer materials on the market call for a new level of abstraction. We are viewing what happens in terms of physics as locked into the biological process.

This spin I speak of, I can get it to go back and forth continually. But you can't do that unless one special product is in the mix-er, RL-37. If RL-37 is added to the blender, a spin one way will deliver an ammonial reading, and a spin another way will yield a nitrate reading.

My formula follows:

Put in water, a humate, calcium hydroxide, magnesium sulfate, Bo-Peep, a special amine compound, castor oil, sodium carbonate and water — it has to be distilled water or good reverse osmosis water — and seaweed extract.

The question is often asked, what is in triple superphosphate that can precipitate out calcium? Triple super is such a powerful cation it attracts calcium so tightly it can't be released. The only way it can be released is to build high enough organic matter with enough bacterial activity to work on it. Eventually bacterial activity will break it down.

As pointed out in *An Acres U.S.A. Primer*, triple superphosphate bonds up quickly and becomes insoluble if organic matter is low. The higher the organic matter, the longer reversion takes. The real trick to using triple superphosphate is good distribution over the acre.

Potash, or potassium, determines the caliber and thickness of the stalk and leaves, the size of fruit and the seed crop a farmer sells. Potassium is required, but not at the levels farmers pour it on. I have a rule for using potassium and phosphate fertilizers. If you finally get a soil in true balance for seed crops, you will want two parts phosphate to one part potassium. For leaf crops and forage — spinach, lettuce, alfalfa — the ratio of potassium should be four to one. Phosphate is a catalyst. It recycles. Potassium is not, and does not recycle. This is proved out when crops are harvested and analyzed. There will be high potassium readings and hardly any reading for phosphate, because the latter recycles. Most of the soils in the upper midwest have tremendous amounts of potassium. It is a negative element, and much of American agriculture uses it for its negative charge, rather than calcium.

Phosphate is always positively charged in terms of physics, yet wet chemistry texts have it show negative. Phosphate will carry an electrical current, calcium won't.

The sources of potassium are many and varied. First, there is potassium sulfate. Here they take potassium chloride, treat it with sulfuric acid or sulfur, drive over the chloride which converts to chlorine–and then they bottle and sell it as clorox. There is some naturally mined potassium sulfate in New Mexico. Generally speaking, that product goes a long way at the rate of 150 to 200 pounds per acre. Sometimes 25 to 50 pounds is more than enough, especially when readied for application in blends. I have it worked out for manual presentation that there is five times more energy in potassium sulfate than there is in muriate of potash. An ordinary conductivity meter will give a higher reading for potassium sulfate than for potassium chloride. There is a very understandable reason for this. The sulfur in the sulfate radical carries a positive charge. Using the Milhouse Unit system, a single molecule of potassium sulfate has 94.750 units, whereas a molecule of potassium chloride has only 19.750 units — a one to five ratio.

A traditional source of potassium is Chilean nitrate of potash. It is a very natural product created by bird guano on native stone. The Chilean government owns this source, and sells a limited amount into the international markets each year. During most wars — WWI and WWII in particular — Chile was a major source of potassium nitrate for gunpowder. The cost is generally high, so it is used mostly for specialty crops.

There is also a product on the market called Arcadian Nitrate of Potash, which is a trade name. Arcadian Nitrate of Potash is made with sodium nitrate plus potassium chloride, the numbers being 15-0-14. By law, they have to state the chlorine content, supposedly.

The next potash source is sawdust, and it has a variable level of potassium. Hardwood ashes are a source of potash, as are tobacco stems, pecan hulls, cottonseed hulls — but if you put cottonseed on the field, look for insect problems. The way to solve that one is to mix cottonseed with tobacco stems. Pecan hulls are good as a fine grind, and they furnish a terrific inventory of trace minerals. They do not break down very fast unless water and bacteria go to work. Straws, rice hulls and organic matter represent potassium sources, but the problem is availability. The trades grind and use rice hulls as carriers in vitamins and trace mineral premixes, and for making vitamin supplements for the human population. Instead of wheat base, they use rice hulls because so many people have an allergic reaction to wheat. Rice hulls used to be a byproduct that could be hardly given away, but during recent years they have fetched a good price on a per ton basis.

It makes little difference whether sawdust is applied kiln dried or wet. However, many soils are so low in basic energy, it takes so much of the soil's energy to break sawdust, the process ends up robbing the crop. Reams always recommended sawdust application at least 90 to 120 days before seeding. That may work in Florida or in sunbelt states, but has to be questioned in northern states, where winter closes down biological activity. Even early fall application would not allow for 120 days of active microbial activity.

There is at least one other potassium nitrate that rates mention here, and that is a 13-0-46 product from Israel. It is reasonably water soluble if you have enough water.

In addition to *Farm Chemical Handbook*, there is a magazine called *Chemical Marketing Reporter*. It lists all the chemicals and prices every week. Once a year they put out a *Buyer's Guide* and list a WATS line number. If you call and say you are a farmer named John Henry, probably no one will deign to talk to you. You have to set up a bank account and create an eye-wash name and tell the banker you are president. When the supplier calls back, the bank reference kicks open the door.

I realize the computations and sources given here walk a country mile beyond the fondest visions of folklore organics. The point I would like to make is that there is a big difference between toxic rescue chemistry on the one hand, and fertilizer energy packages that can be computed and harnessed for crop production in such a way they make crop protection chemicals unnecessary.

I do not know whether I can resolve the debate between so-called chemical agriculture and those who reject all man-made compounded fertilizers as inappropriate. I do wonder whether the debate should not be more between the elements of nature and man-made molecules — in which case I might tend to side with re-jection of chemicals of organic synthesis. After all, their vibrations were not there when the command was issued, "Let there be light."

12

FEED THE SOIL

Nature's way is largely biological. It combines both chemical and organic systems. The bird that leaves its dropping on a calcium rich seashore ledge sets up an acid reaction for crop fertility just as surely as the chemist who apes nature's system in the laboratory and in the factory. We know that life cannot endure unless nature is allowed to produce nutrients for the biological systems of the soil for plants that nourish therefrom and are then consumed by animals and man. This is possible today only if man, in his ignorance, does not tamper with either the organic or chemical systems. Our vision of matter's division into anions and cations has provided the key. Now we need only the wit to use that key.

There was a paragraph in an early issue of *Acres U.S.A.* that merits republication. Crafted by the late C.J. Fenzau, it declared as early as 1972 that at least one publication was on track when the organic movement gathered speed for its sweep into the present era.

A soil system for nutrient and energy production is a living system in which bacteria and the many soil organisms must receive nutrition and energy from proteins, carbohydrates and cellulose and lignin — all organic materials — in a soil that has a managed supply of both air and water within a balanced chemical environment. This chemical involves more than just N, P and K. It requires an equilibrium of pH, calcium, magnesium, sodium, potash, humus and a nutritional balance of sulfur to nitrogen, nitrogen to calcium, calcium to magnesium, magnesium to potash and sodium. Without this balanced equilibrium, neither the organic system nor the N, P and K system has any enduring potential for soil building or plant nutrition. This total equilibrium is even more essential for the vital nourishment of man.

On the other hand, hydroponics is not a living system of plant nutrition. It is a synthesized method of supplying elements for plants growing in an environment protected artificially from the hazards and basic elements of life itself. The time clock and configurations of life are continually influenced by hormone-enzyme systems, by light, temperature, carbon dioxide, photosynthesis, antibiotics and a balanced viable plant sap system which requires basic hormones and a specific pH equilibrium in order to biologically produce a healthy plant. This plant must grow nutritionally, ripen and dry. The nutritionally ripe plant is essential for complete animal or human nutrition. Neither the chemical nor the organic system can assure fully matured or "biologically ripe" food production. By coincidental accident, our agricultural system can produce the desired food quality without the hazards of weather, insects, disease or geographical limitations — and hydroponically, we could be a nation stuffed with partially matured carbon-based compounds that look and feel like the food. In reality this so-called food results in partial or complete malnourishment — the predecessor to poor health and eventual metabolic and physiological deterioration.

Perhaps it is unfortunate, but nevertheless essential, that *we must farm the soil*. It is only folly that we invite the hazards of hydroponics to sustain a synthetic system of life. Since we must cultivate soil for production of both nutrients and energy as well as fiber, we must begin to feed the soil, not the plant. This is the basic and fundamental change required for progress to begin.

This passage in effect points out that farmers were beginning to express their own experiences as early as 1972, and this was irritating authorities. Many had already come to regard nitrogen as a source of nutrition and energy for the actinomycetes fungi to digest and break down the complex carbons contained in lignited organic matter. Many looked to the day they could make their natural nitrogen and carbon cycles work. In the process they reached for new tools and ideas which the authorities hardly knew existed. The reporters on the metro dailies and those serving industrially owned farm magazines — taking their points of view from the university — slowly lost touch with the countryside and the challenges its free spirits were issuing.

The Reams system has never made it into the texts, but its genius is being translated into improved crop production. Many of the fertilizers he learned to harness can raise the crop brix level, proving that, indeed, the anatomy of insect control is seated in this brand new form of energy management.

In countless one-on-one Socrates sessions with Reams, I picked up on many of the facts I now pass on to farmers. The grand lesson is to always experiment — and compute. Whenever there are questions, there are also answers — if we have the patience to find them.

One question that often comes up at seminars is, *How does potassium hydroxide compare to potassium sulfate in energy?* It can be calculated, of course, and the arithmetic says it is about the same, albeit not quite as high. The potassium in potassium hydroxide is 69%.

What if the soil has 400 to 500 pounds potassium, based on a water-soluble test? How can the proper phosphate-potassium ratio be achieved? One quick way is to add a little calcium nitrate in liquid form, and the potassium ratio will drop. Lime will also bring it down.

Nitrogen is the major electrolyte in the soil, and in order to have our solar system we have to have the sun. For any plant or animal, or the center of a cell, the force that draws other elements to it and starts the basic magnetism to "plate" and make a new cell is nitrogen. No matter what kind of cropping program a farmer plans, he has to account for a nitrogen source. The amounts vary, as does the contribution of the natural nitrogen cycle. We quite frequently come up a little short on nitrogen. There may be enough nitrogen to carry the day, but not for the growth phase or total crop sequence. There is a reason for this. The or-

THE ORP INSTRUMENT

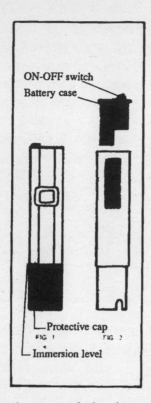

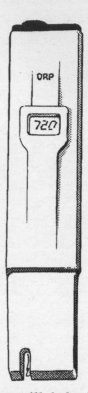

ON-OFF switch

Battery case

Protective cap

Immersion level

This is one of the instruments that will help farmers take mainline farming into Century 21. For details, see the next pages. Operating instructions are simple in the extreme. To check ORP, dip it in Redox solution (HI 7020). The reading should be 200-250 mV at 20 degrees C. Otherwise clean the electrode by rubbing down the probe with a soft cotton, soaked with a cleaning solution (HI 7061). Always use rubber gloves.

REDOX VALUE

Oxidation reduction potential (ORP) reactions in the soil are of major significance both from the chemical as well as the biological standpoint. Since an electron flow is involved, a redox reaction produces an electric potential, which is usually measured with a platinum electrode versus another solution contacting reference electrode.

Since ions are involved, the ORP and pH are measured in a water solution. Life obtains its energy from the oxidation of reduced materials. In soil electrons are continually being transferred by the biological oxidation of organic matter. In soil organic matter is the major electron donor. Oxygen is the major electron acceptor.

Oxygen is the only electron acceptor which plant roots can utilize. Respiring life captures energy from the transfer of electrons in a series of reactions involved in the movement of electrons to oxygen. In those cases where oxygen is not available: nitrate; ferric (Fe + 3) [fully oxidized form of iron]; Mn + 2; and sulfate can act as electron acceptors if the live organism has the correct enzyme systems.

Electron flow is reduced as organic matter is converted to humus and increases as fresh organic matter is added. Reversible reactions produce an initial fast responding ORP, whereas irreversible reactions result in a slow drift. In those reversible reactions where equilibrium is reached, electron transfers cease. Hence a stable ORP is read with zero current flowing through the electrodes. Soil systems are typically irreversible, especially biological reactions.

Like pH, the redox potential does not affect plant growth directly, but may control influential environmental conditions. The ORP tester is the simplest instrument for measuring the state of soil oxidation. The ORP value will drop due to spreading of manure or organic matter and therefore planting may be delayed to allow the potential to return to the former approximate value. ORP at this time is perhaps used most to determine compost stability.

For neutral or alkaline pH soils, the availability of phosphorus, iron and manganese may be made more available by not maintaining the rH value at too high a level. It may be better to have an rH of 26 vs. an rH of 30. — *Pike Lab Supplies, Inc., Strong, Maine.*

ganic matter complex in most soils will not permit as much release as expected.

When I plan a corn program — and I am using a 28% nitrogen, or ammonium sulfate — I have a tendency to bump it up and put on more than indicated.

Needless to say, any discussion on fertilizer materials could be fleshed out to fill a chapter for each item. Up to a point we could put it all together and get crop production underway, in a way exhibiting an arrogance no mortal should be guilty of harboring. And up to the end of WWII, some justification attended our learned arrogance. And then came the discovery of the DNA and RNA by James D. Watson and Francis Crick.

It was known, of course, that radiation affected the mutation and death of cells long before a nuclear device was detonated in the New Mexico desert. It was also known that the ionized chemicals used in agriculture had somewhat the same effect on cells of plants, animals and man. Not known was how and why. The birth of the atomic age somehow unfroze funds for the appropriate research, a story that has been related to the world in that scandalous book, *The Double Helix*, by James D. Watson.

In discovering DNA, Watson and Crick kicked open a door to understanding second only to the energy equations of Einstein and Milhouse. Using only a short code, the DNA stairway in the chromosomes of a simple cell might be 1/2,000th of an inch long and contain 170,000 steps. A DNA might be no more than 1/300th of an inch long and hold several million steps. Andrew Zaderej, a Ukrainian scientist transplanted to the United States, has computed that a single human cell contains as much information as several sets of encyclopedias — with this difference: each bit of information in a cell is important, even vital.

The gateway to the cell is about 100,000th part of a millimeter thick. Through this membrane must arrive the nutrients that say grow or die. Particles of nutrients have to cross over on their own stream. Obviously ionized chemicals called fungicides and insecticides have their own key to cellular entry.

A cell has to have nitrogen as a core to start up the magnetism necessary to build another cell in a plant. The next requirement —

carbon, oxygen, hydrogen and calcium. Four of these elements can come from the air, but not calcium. To some degree the farmer can put things together, and then the code takes over. The DNA determines the kind, species, color and identity of the plant. From that point on, one key element has to come into play, namely the catalyst phosphate. For the average plant, over 85% of the nutrient load can come from the air. In fact most of any plant's nutrition comes in through the air in the form of carbon dioxide — and, of course, some nitrogen. Air is approximately 78% nitrogen.

Calcium is the key ingredient in the plant charged with the task of drawing nutrients from the air. Almost all upper midwest soils have ample calcium, the trick being to get it into the plant. It should now be at once apparent that planning a fertilizer program is like playing pool. Sometimes it is necessary to bounce off two other balls to put the right one in the corner pocket. In many cases no more calcium is needed in the soil. Needed is a way to bump it into the plant. There are times when it is possible to increase calcium uptake by increasing the amount of some other element in the soil. Moving west from Minnesota, for instance, there is often enough calcium in soils, so it becomes a bumping game. In the eastern part of the country, that isn't quite true. There are many soils that are simply deficient in calcium.

I have found that the use of calcium nitrate really works quite well in some situations for this business of bumping calcium uptake into the plant. Once I get the ball rolling, it rolls pretty well.

I have stated it before, but this one point requires routine reiteration. Phosphate is a catalyst. It is also the key ingredient in engineering the formation of sugars in the plant. A farmer has to understand this so that when he looks at a plant he can recognize that when certain nutrients are short, it may be because phosphate is short.

Getting too much calcium into a plant is not very likely because there won't be energy enough to carry it in. And, of course, the primary fertility element needed to carry it in without phosphate is nitrogen. Nitrogen can carry all the basic elements into a plant. It takes a lot more water — when there is not enough phosphate — to grow a plant if nitrogen is handling the carrying chore, than it does if an adequate phosphate load is present.

The conventional wisdom in conventional agriculture has nitrogen carrying potassium in so a plant can be grown with potassium, nitrogen, and a nominal amount of phosphorus. This produces a plant low in sugar and subject to harvest shrink. And, as we have seen, a plant with a low brix reading is forever subject to bacterial, fungal and insect attack. Finally, this conventional practice has become an insurance policy for the sale of lots of rescue chemistry. A plant must have calcium.

All forms of manure are cationic without exception. This includes composts.

Ammonium sulfate is one product that rates special attention because it works well with the livestock in the soil. If nature heats it, it freezes. If the weather freezes it, it heats up. It should never be used in low calcium soils. Ammonium sulfate has a double charge. Both the sulfate and the ammonia are positively charged, and low calcium will knock the aluminum off the clay molecule—which is standard textbook fare. When you apply ammonium sulfate, mix in about 15 to 20 pounds of dry humates. The products magnetically attract each other. The best source of ammonium sulfate looks like it is dirty. It has a dark color to it. For optimum temperature control, approximately 200 pounds per acre is recommended. Normally you don't put much more than that on an acre at any one time. In the heat of August, it will generally maintain your soil temperature around 72 F. which is the optimum for maximum microbial activity, even though the air temperature is 100 F. or more.

How long will it last in the soil? For a while, you may need to put it on every year, but as you build an ammonia reserve in the soil, the soil will automatically do this without any ammonium sulfate.

I was going to a veterinary seminar in Kansas City, Missouri a few years ago in March. It had just snowed. I got on this little twin prop plane that flapped its way to Kansas City. On the way I was looking out the window and every once in a while there would be a field where the snow had all melted off. It was a perfectly black square with white all around it. One of my colleagues was along with me and I said, *What do you think about that down there?* He said, *So what?* I said, *How did that happen down there?* He looked down and saw a guy hauling manure. He said, *Oh, he just spread manure over that area.* I was just fascinated. It convinced me then that there had to be a difference in soil temperature even during a snow time of the year. The same is true for good compost. Can you apply it in the fall without loss? Yes, and you can apply it in the spring as soon as the snow is off the ground. There doesn't seem to be a lot of loss to the air with it. It seems to go down. Does it help to keep the insect population down? Yes! The reason is that the crop isn't under stress at 72 F. in the soil, and when it gets to 90 F. in the soil, and the crop is under stress, the bugs move in.

A high aluminum uptake sets up all types of strange things. It stunts plants, then shrivels them. Under aluminum assault, seeds may not even sprout. These anomalies may not be at once apparent, for which reason the mischief is deferred until animals are fed. A high aluminum concentration will affect the central nervous system. If recognized in time, calcium can be used to counteract the effect. There is a product put out by Eli Lilly of calcium gluconate with vitamin D that is excellent.

There are products that should be avoided, except perhaps for golf course application. Milorganite is one. Milorganite is nothing more than dehydrated sewage sludge from the city of Milwaukee's sewers. In some cities — Chicago included — they have stated right on the bag not to use the product on soils that grow food for human consumption. Too many chemical plants sneak their toxic materials into the sewers to make farm use for these materials feasible.

Sewage sludge from any town — even one without industry — is inherently dangerous. Even in small communities, household cleaners enter the pipes and end up in the settling ponds. Lime is almost alone as an antidote for such toxins.

Chicken manure has a high value in terms of numbers, but is not available in all parts of the country now that poultry and egg production have localized chiefly in the east and the south. Too much chicken manure can burn a crop severely. It is absolutely mandatory to check the material for how many pounds or gallons per acre the soil can handle. Poultry manure from caged layers (not litter) has lots of trace minerals, but it is low in carbon content. This means application to a low organic matter soil could worsen the situation even though there are a lot of bacteria in the manure. Bacteria without food soon die out.

Dried blood used to be a good soil amendment, but they process it now and put it in livestock feed and get a lot more money for it. The economics of fish meal largely has deleted that fertility assist from the farm scene. Sodium nitrate isn't used too often anymore. It is used more in the food industry and the price has taken it out of the marketplace. It is a negatively charged element. It would prove useful on lettuce, celery, spinach and cabbage crops. The amount would have to be based on analysis, but in some situations 300 or 400 pounds per acre might be indicated. The problem you run into is sodium, and the only way to counteract sodium is to add back plenty of compost, or Z-Hume, a liquid humate product with enzymes added.

Ammonium nitrate is a "double one." The ammonia is positive and the nitrate is negative. Normally, this cannot be warehoused in the same bag, but by coating ammonia with diatomaceous earth, leaving the nitrate uncoated, the problem is solved. If you put it on corn, the nitrate goes to work first and produces growth. About the time the nitrate wears off, the ammonia takes over and the seed crop arrives. Thio-Sul is 12-0-0-26, and 26 is for sulfate. It is a double cationic substance and is very powerful. When there is trouble getting seed to set, that is the role for Thio-Sul. The tomato is a common plant with the problem. It is not uncommon to get a nice big plant with lots of leaves, and that's the end. Nothing happens. On a garden basis, you put two cups of vinegar in five gallons of water and sprinkle around the plants and they'll set down fruit like mad. On a field basis, Thio-Sul is an excellent product to get enough of a cationic charge to kick everything into seed production. It creates enough energy for the plant to pick up manganese. These are slightly to the positive side, but they have a lot of negative in them as far as energy charge is concerned. It depends a little on how they are manufactured. Chiefly they are made with ammonium nitrate plus urea. Some places use ammonium nitrate — sodium nitrate plus ammonia. There's a new one on the market that is 40% and water soluble. It is a blend of several nitrogens together and it has a tendency to be primarily on the positive side.

We should mention also aqua ammonia in this brief survey. In some areas of the country, elevators will buy anhydrous and they have a converter to run anhydrous through water to make what we call aqua ammonia. About the most that they can get the water to hold — depending on the facility and the type of water — is from 23% to 28% under certain conditions. Some of the larger farming op-erations have their own converters and make their own, adding some of the water soluble products available. They also add molasses or sugar or humates to the mixture. One comment that I want to make about urea is putting it on in dry form on an acre of land has a tendency to be quite harsh, and it has a dehydrating effect. If you are getting plenty of rain, it won't make much difference, but if rainfall is marginal it will deliver the net effect of a loss of carbon because of the excess heat it creates. It is very hydroscopic. It draws moisture rapidly.

Iron is a very important element in the leaves of any growing crop for the purpose of drawing more energy in — the anionic rays from the sun. The rate of photosynthesis is vastly affected by iron. If there is adequate iron, leaves will always be thicker. If you have a one inch and two inch pipe, you can get four times more volume through the two inch than the one inch pipe. If the thickness of a leaf is increased from 1/64 to 1/32, the potential factory capacity has been increased accordingly. What's more important, to have a bigger leaf that's thinner or a smaller leaf that's thicker? Thickness is more important than size. When you see this you know that the plant has gotten the right nutrient ration.

I realize many of the statements made here are a tough bullet to bite. The organic folks may well be right, that the man-made fertilizers are inherently dangerous. They are dangerous, and they are not playthings for amateurs. It takes judgment and maturity to bring this science into harmony with the living system of the soil.

This cannot be done by chemical amateurs who rely more on crop residue for a buffering agent than on meeting nature's requirement in the first place.

Speaking of a living system and crop residue, there are two other products on the market that are totally organic that are doing a great job on crops. They are called Arouse and Crescendo.

Arouse is a soil product that is placed with the seed at planting time to enhance root development and increase humus and nitrogen in the root zone. It contains a broad spectrum of specially selected bacteria chosen for their ability to enhance the root environment of plants, ultimately making necessary nutrients more available for plant assimilation. Once these special bacteria establish a culture in the root zone of commonly grown row crops, the result is an unfriendly environment for many common row crop root diseases. Root rot, nematodes, maggots and root worms, all are problems that noticeably subside once the bacteria culture is established. It is unique in that all of the nutrients that are necessary to establish the specialized bacteria are included in the product. This is a common problem with many bacteria products on the market today. Many times very good bacterial products are applied to the soil only to find a very hostile environment, such as lack of nutrient, air or water, which makes it practically impossible for them to establish.

The method of application is also designed to give the bacteria a better chance for establishing in the soil. Always remember that the aerobic bacteria needs nutirent, air and moisture. Since this product contains the nutrient needed to get the bacteria started, all we need is air and moisture — and no direct sunlight. If it is applied in the furrow with the seed, it should be in an environment that has moisture, air and no sunlight. If the seed can germinate, the Arouse bacteria should be able to establish. Once the bacteria have established and have depleted the supplied nutrients, the nutrient is provided by the plant rootlets throughout the remainder of the growing season. All plant root systems have a base exchange, and as the old rootlets drops off and new ones establish they supply nutrient for the bacteria introduced at planting time. This rootlet residue is rapidly converted to humus and humic acids which are powerful chelating agents and help the plant acquire plant foods more readily. The production of humus by the Arouse bacteria in the root zone increases the water-holding capacity of the soil, thus improving the ability to fight weather stress later in the season; at the same time that humus is being produced, nitrate nitrogen is also being produced. The increase in corn yields expected from using Arouse can be anywhere from six to 12 bu./acre, also grain quality has been affected quite dramatically. An increase in protein by 1% to 4% and sugar levels increased from 10% to 20% more than controls can also be expected.

Crescendo is a sister product to Arouse with the same bacteria package but manufactured into a dry soluble powder so it can be mixed with water or starter fertilizer for application. One incident I would like to tell about is where we had a 60 bu./acre increase of corn with Crescendo over the control. The producer was applying 10-34-0 starter on the seed at planting time. Applying liquid fertilizer on the seed is never recommended by me, but this is what was being done. When he started his test plot he just added Crescendo to the 10-34-0 for a few rounds, then a few rounds without, etc. When he combined the corn he was in for a big surprise. Wherever he added the product to the starter he had about 10,000 more plants/acre, therefore the large increase in yield. The bacteria had set up an environment in the seed zone so that the corn could germinate. This is a good example of how damaging chemical fertilizers can be if used in the wrong way, but by adding a biological product to it made a large difference in the crop.

All of the materials mentioned here can raise the brix levels of crops. There is not one chemical of organic synthesis — pesticide, fungicide, herbicide — that can raise anything even one brix point, and therein lies a distinction. This reality reconfirms the proposition that insect and weed control are seated in fertility management, and not in using more powerful goodies from the devil's pantry.

13

A MIX OF COLOR MOLECULES

As we waltz our way through the many lessons Carey Reams left behind, the shoulders he stood on keep coming into view. For instance, the forgotten scientists who first worked out the colormetric test as a measure of dry ions rate a mention. They discerned that a certain volume created a certain density of color. They tested nitrogen in the soil and in urine within the parameters of this theory. They developed the test that linked nitrate nitrogen to the color blue, a test now part of the standard LaMotte procedure. They were able to do this because they figured out how many molecules it took to provide density. The computations by Reams confirm the entire procedure, and it is the colormetric concept that now stands ready in the wings to serve a more scientific mainline agriculture.

Many of our farmers have come to some of the same conclusions pragmatically. They have correlated the color green with an appropriate level of nitrogen, and a mix of color molecules with nutrients out of the equilibrium needed for proper plant growth. When a field has a metallic sheen, the crop will be healthy. On small grain, a golden color is something devoutly to be wished. It isn't seen very often, but when it shows up it serves up some real excitement.

Iron helps a leaf hold more heat, and this means the rate of photosynthesis occurs more rapidly, and more sugars are produced. When enough sugars are produced, the plant in turn produces more oils. When the oil content of a crop is increased, shelf life has been enhanced. Sources of iron are soft rock phosphate, 4%; basic slag, 40%; and iron sulfate, 32%. The simplest way to check the benefit of iron is to introduce iron sulfate into the soil and check the results. In most cases the limiting factor for carrying iron into the plant is phosphate and humus. When looking at a field, it is well to remember that the problem may be the catalyst phosphate — or nitrogen — and not iron.

Someone has said that the real energy in agriculture is knowledge. Knowledge contains photons never fully comprehended — in fact, scientists have gotten off the track because they have never backed up all the way to *cause. All energy comes from the sun* is more than an axiom. It is the foundation for science and for soil science.

Two or more protons make an electron.

Two or more electrons make an atom.

Two or more atoms make an element.

Two or more elements make a compound.

Two or more compounds make a substance.

Two or more substances make a cell.

Two or more cells make an organ.

Two or more organs make a system.

Two or more systems make up bodies, human, animal or plant.

The problem with too much science is that practitioners have no interest in the results of research that cannot be patented for profit beyond the dreams of avarice.

The April, 1990 issue of *Acres U.S.A.* carried a report by Arden Andersen which was published without indecent profits in mind. In it he talked about energetics filtering into the ranks of production agriculture. This report merits inclusion here as an abstract in depth.

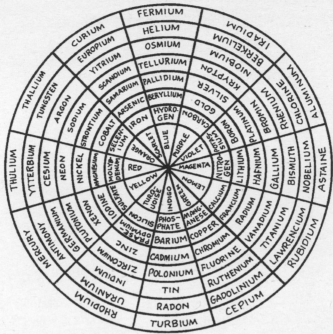

Energetics has received much interest relative to energy-field medicine. Most noted in this work are Albert Abrams, Georges Lakhovshy, Vlail Kaznacheyev (who demonstrated that diseases could be induced or reversed with electromagnetics), Robert Becker, John Ott, Bjorn Noordenstrong, John Zimmerman, and Fritz-Albert Popp. The most noted work in agricultural energetics has been done by Philip Callahan. Energetics adds to agriculture the causative aspects of soil, plant, and animal nutrition. It presents a view of the innermost secrets of nature. It teaches us to be better observers and to use our common sense. As Phil Callahan and Popp have shown, natural phenomena are primarily energetic and secondarily chemical/physical. In other words, the energy fields of organisms and chemicals interact first.

This interaction results in the chemical-physical phenomena we observe. Consequently, we can evaluate these energy fields to arrive at a truer picture of what is actually happening. When we combine these data with the chemical test data, some understanding of both, and a grain of common sense, we can solve almost every problem we face in soil and plant nutrition.

We know from Callahan's work that *form* and *frequency* go together hand-in-glove. As such, the shape of something affects its activity and

MAGNETIC SUSCEPTIBILITY SCALE

	MAGNESIUM SUSCEPTIBILITY (cgs)
Zinc	-11.4
Zinc Carbonate	-34
$ZnSO_47H_2O$	-143
$ZnSO_4$	-45
KCl	-39
KoH	-22
KNO_3	-33.7
K_2SO_4	-67
K_2CO_3	-59
$FeCO_3$	+11,300
$FeSO_4$	+10,200
$CuSO_4$	+1,330
CuCl	-40
MnO	+4,850
$MnSO_4$	+14,200
$MnCo_2$	+11,400
$CaCO_3$	-38.2
$Ca(NO_3)2$	-45.9
$CaSO_4$	-49.7
$CaSO_42H_2O$	-74
CaCl	-54
Ca	+40
NaCl	-30.3
Na_2CO_3	-4
Si	-3.9
SiO_2	-29.6
Sm	+1,860
Sm_2O_3	+2,282

relationship in any biological system. This can readily be seen in taking magnetic susceptibility as a measure of a material's antenna characteristic. Chemical tests do not distinguish between sources of the same nutrient. Nitrogen, for example, is supplied chemically in ammonia, nitrate nitrogen, and urea. Each source has a different shape and a different magnetic susceptibility. Ammonia has a tetrahedral shape and a magnetic susceptibility value of -17 centimeter grams per second. Nitrate nitrogen has a flat, triangular shape and a magnetic susceptibility value of -33.6 centimeter grams per second in the form of ammonium nitrate. Urea has a flat, triangular shape with a "handle" on it and a magnetic susceptibility value of -20. Almost every element varies as it is combined in different forms, as shown on the accompanying scale.

Highly fertile soils have positive magnetic susceptibility values and are said to be *paramagnetic*. Sterile soils have negative magnetic susceptibility values and are said to be *diamagnetic*. The fact that a soil is highly paramagnetic does not guarantee high fertility, but it does indicate high potential fertility. As can be seen in the preceding brief list, the rare earth metals and the common trace metals have high paramagnetism. These are all important for balanced nutrition and are generally present in mineral powders, rock dusts, soft rock phosphate, and so on. As can be seen, the types of fertilizer materials used affect the soil energetically as well as chemically. This lends importance to the evaluation of soils and plants energetically. It allows for the observation of causal factors in nutrient balancing and the effect of various nutrient compounds. As a result, many nontraditional fertilizer materials have been discovered to be vital to soil regeneration and plant feeding. They include vitamins like B-12 and C; sugars like molasses, sucrose, and dextrose; trace elements like silicon and iodine; and even colors.

In his book *Light, Radiation, and You*, John Ott documented his discovery that plant functions and maladies can be affected by simply altering the color of light irradiating them. Research has shown that there are times when success or failure of a fertilizer can be traced to the color of the dye used in the fertilizer. This is not necessarily a major or primary consideration in formulating a fertilizer, but it may be a factor in solving a peculiarly difficult problem.

In nature, colors are an integral component of bioelectromagnetics. The various colors of flowers act to "pump" the nectar radiations, thus intensifying the signals, which insects, particularly bees, and hummingbirds home in on. This is perfect because most nectar feeders are day feeders and the prominent background radiation is visual light. Night feeders, on the other hand, would not be aided by visual radiation pumps because the prominent background radiation at night is infrared. Colors can also act as energy "pumps" in fertilizers, as well as being independent photon stimuli, as Warren Hamerman points out.

It is interesting to note that though the flower color is important for intensifying the signal attracting feeders, it is not the message carrier. The signal itself is the message that the feeder interprets to decide if food is present. For example, it has been shown that if the nectar drops below a refractometer reading of 7 brix, honey bees will not feed on it. They will not even be attracted to the flower. There simply isn't enough nutrition present for the bees to feed on. This is a very important point to make because proper fertilization can and does raise the nutrient density of flower nectar, as well as the entire plant. Traditional agriculture totally ignores this possibility in favor of spending millions of dollars for genetic research which never solves the real problem: *nutrition!*

Bioelectromagnetic "pumps" are important to many aspects of nature. It has been shown by Philip Callahan that ammonia in the soil and plants acts as a very strong "pump" for radiations tuned to by insects. In other words, when there is an excess of ammonia present in a field of corn, the signal radiating from this field for insects to pick up is intensified. This of course makes it much easier for the insects to home in on the corn field in time for lunch. The question must be asked, "Why would there be excess ammonia in the corn field?" The answer is very simple. Modern agricultural practices result in the soil biosphere becoming anaerobic (lacking oxygen). Organic materials are then not converted to humus, which requires oxygen, but rather to organic compounds like alcohols, aldehydes, and preserved organic materials plus ammonias. This phenomenon results in a sterilizing of the soil-biological system which results in inadequate nutrient interchange between plant and soil. This in turn results in less than optimum plant integrity, health, and nutrient density. The radiation pattern emitted by the soil and plants reflects this state of integrity, exactly. The ammonia seeps into the air very effectively "pumping" this radiation. The signal to nature's "garbage collectors" is, "come clean up."

Shown with this report is a color/element wheel some colleagues and I [Arden Andersen] formulated a couple of years ago. It is a guideline for study and experimentation. There are some similarities to Dinshah's work, yet the differences are intentional. Those familiar with Carey Reams' teachings will notice a total of 84 elements on this chart may be varied or rearranged slightly.

It is interesting that biological systems will balance from the inside out. The same holds true for this chart. If the inner ring of twelve elements is in perfect balance, the remaining 72 elements will automatically fall into balance. The reverse does not seem to be true. There also seems to be considerable correlation to opposites, such as iron, manganese, and copper; carbon, silicon, and zinc; nitrogen, molybdenum, and magnesium; and so on. An interesting observation occurs in the yellow and violet rays. Traditional fertilization has overburdened our soils with potassium chloride. Our waters are polluted with chlorine. Our mouths are filled with mercury. Radioactive plutonium is mass-produced for weapons and fuels. Most city street lights are sodium vapor, which predominantly radiates the yellow wavelengths. All of this possibly allows for aluminum to precipitate in the brain, which is correlated to Alzheimer's disease. In color therapy, yellow is often correlated to intellect and thinking, whereas violet is correlated to cleansing and purification. It is interesting that national student test scores have declined and that there seems to be an increase in aberrant behavior. Perhaps these are all coincidental correlations. Perhaps not. In color therapy it seems that colors will aid in assimilation of the opposite element, *e.g.*, orange for calcium. It can also be noted that oxygen is not listed on this chart. The reason is that oxygen is inherent in almost all of the elements listed when they are involved in a biological system. Therefore, it did not appear as a separate element.

It is important to keep in mind that this or any nutrient chart is quite mechanical relative to the actual system of nature. Manipulation of elements in nature is not as cut and dried as it is on the chart. Nature is a dynamic system, a whole comprising more than the sum of its parts. The addition of the color wheel partially addresses this issue. As can be seen, any time a color and/or element is addressed, an associated vitamin will also inherently be addressed. We could then correlate this to nucleic acid, amino acids, proteins, tissues, plant parts, and so on.

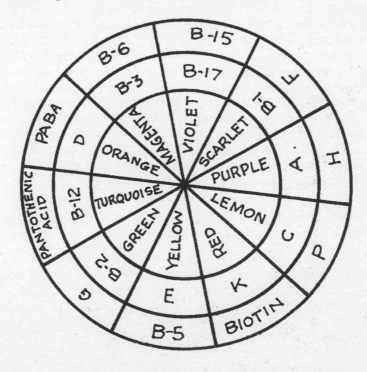

[Implied also is the proposition that the dye green added to a fertilizer can repair the manganese problem, general balance for copper, chromium, fluorine, ruthenium, gadolinum and cepium to follow if all the first ring nutrients are balanced.]

You will notice that the vitamin chart does not have the indigo-blue rays. My reason for this is that in biochemistry we learn that in all functions of the body the energy-transport system of adenosine triphosphate is involved, and, therefore, the phosphate-hydrogen "rays" of all the charts, whether mineral, vitamin, amino, acid, etc., would be involved as phosphate-hydrogen rays. In other words, they do not correspond to any specific vitamin but rather to the energy-transport system of the whole biological system. Comments, observations, and suggestions are appreciated.

The color/element wheel works very well for explaining how insects home in on deficient plants. When a plant is nutritionally balanced, its energy field radiation will be as white light. White light is a collection of all the colors equally. When a plant is deficient in a nutrient, say iron, there is a deficiency in the corresponding color (scarlet for iron) component of the plant's energy field. This, in effect, produces an energy field prominent in the opposite (green opposite scarlet) component. Consequently, the insect interprets this as a predominant source of manganese. The insect homes in on this and attacks. A balanced plant does not broadcast any color prominence, just white light, which the insect's antenna is not designed to receive. In this way, nature purges the food chain of deficient foodstuffs and maintains healthy consumers on up the food chain. It is truly perfect divine order.

The study of energetics is providing insights into solving the problems of insect and disease infestations, crop quality and yield, and environmental integrity. How wonderful simply to balance the soil and crop nutrients, consequently the animal and human diet, and solve the problems over which genetic engineers will be enabled to pursue grander accomplishments.

As more scientists, consultants, and farmers share their insights, we can collectively solve the impending environmental, health, and economic problems of America and the world.

So ends Arden Andersen's brief essay on light and color energetics. We have still to integrate what we know, knowledge being boarded in air-tight compartments as a matter of habit. Colors hint, and nutrients abide. This business of adequate phosphate being more a problem than iron isn't limited to field crops. This lack of a phosphate-iron connection often restricts pin oak growth. Iron has a relatively high atomic weight. This means it has a high energy requirement to be magnetized into the plant. Iron helps deliver a dark green color — not the position opposite manganese — but this could be a deficiency of nitrogen and phosphate first. Nitrogen carries the current to transport iron into the rootlet. Phosphate helps convert it into a form that can be taken into the plant. Thus it is possible to potentially have a shortage of nitrogen and phosphate before there is a real shortage of iron.

Copper — or the lack thereof — is most frequently noted when fruit trees do not produce. They do not produce because the bark cannot stretch. When the bark cannot stretch, sap can't flow. This situation can be remedied at times by applying copper sulfate, but many times that device will not work. Again, a nitrogen or phosphate deficiency might be identified as the cause. In other words, there may be enough phosphate to accommodate the basic functions of the plant, but not enough to handle copper and iron needed from a standpoint of energy. There are a couple of products on the market that might be helpful. One is Sul-Po-Mag. It contains sulfur, potassium and magnesium, and it makes copper available to the plant.

The bark on trees is sometimes so tight that four and five foot long cracks develop. Often the bark explodes and pops off. One farmer from New York told me he thought his neighbor was shooting at him. He examined the tree and saw the bark completely split off, so he figured the tree was wasted, and cut it down.

When Carey Reams first encountered this phenomenon in an orange grove, he tried the Sul-Po-Mag remedy and ended up in court. He had to delay going into court as long as possible, and he managed to put six months between the filing of the suit and a day before the judge. When the time came he required the jury to view the facts in the field. When they arrived at the orange grove, not a thing was wrong. All the trees had healed over during the intervening six months.

Some growers have trees that will not set fruit. A lack of sap flow is generally the problem. When an inadequate supply of sap flows up under the bark, nutrients are not in sufficient supply to make fruit. In the process of photosynthesis, sap flows down the center, but just under the bark is where it rises to service branches and fruit.

The normal rate of application for Sul-Po-Mag is 200 pounds per acre every ten years. In the North Temperate Zone, the most effective time to apply Sul-Po-Mag is from July 15 through September 15, if the application is via the soil. In Australia, the season would be the same, but the date would be January 15 through March 15. On a small scale, the application would be about five pounds per 1,000 square feet. This product is often sold under the trade name Dynamate. Dynamate is also an excellent product to put into hog feed for tail biting prevention.

I have been on farms that looked like a rootworm convention. I ran the gamut trying to figure out a remedy other than toxic chemicals. I finally got to Dynamate, and the energy indicators took off. I figure it made copper available and the plant healthy enough to ward off insect attack.

I have had a lot of farms spray on four pounds of copper sulfate to the acre in 20 gallons of water after the crop came off in the fall. As a consequence, these farmers are having a lot less mold troubles the following year. Plants, generally, become susceptible to molds because of stress. This stress might be nothing more than high humidity and a lack of air flow.

Boron deficiency is of the same genre as copper deficiency. Boron causes hollow heart. Calcium excess causes scab. Most of the time trouble starts with a phosphate deficiency, and possibly a nitrate and calcium problem. If these three precursor problems can be straightened out, there isn't a boron problem. The real boron problem with potatoes is a failure to put on calcium, the shortage of which sets up potato scab. The remedy here is to put calcium on the potato acres, and then add something acid enough to foreclose potato scab getting started. Most of the poultry manures are excellent boron sources. So are horse stable manures. And the chemical companies have their armamentariums. Salesmen like to move in like the Vandals who sacked Rome when they find hollow alfalfa or some other boron shortage indicator. Hollow stem is the favorite indicator, not only for alfalfa, but also for cauliflower and broccoli. This may or may not be the case, there being so many other possible deficiencies, any one of which can affect the color. I have seen them sell boron as high as $15 per pound. Yet one can go down to the grocery store and get 20 Mule Team Borax for about 50 cents a pound. Other sources are seaweed and liquid fish fertilizers. Boron helps harden the cell wall to give the plant more strength. Boron is a natural germicide. It is used generally as a wash disinfectant.

Manganese is a first circle element. The key thing about manganese is that if it isn't present in the seed, sprouting simply will not take place. It has a role in the scheme of things that may never be understood entirely. All we know for certain is that when manganese is activated, it causes the DNA to take over and design the plant according to that very complicated set of blueprints in the code. Good sources of manganese are Manzate and Maneb, both are common garden supply house products. Landscapers often use these products to control lawn diseases. Sawdust, pecan hulls, rice hulls, fish and some few other recycleable products are good sources.

Parenthetically, it might be well to point out that one of the biggest problems with pecan groves is to keep them producing and not falter in yielding crops, the latter forcing removal of the grove. The biggest missing link is not spraying the pecan trees with manganese on a regular basis. Oftentimes pecan trees are also short of phosphate. Manganese does more than trigger the sprouting of a plant. In pecan production, it's the total fill of the nut. When the stone of a plant such as the apricot and the peach is empty, this is a good indication that there is a lack of phosphate or manganese.

Soft rock phosphate has a key role in keeping manganese available. It should be put down first — but not in the soil — dressed with compost and manure. Rain should be allowed to carry the nutrients into the soil. This will build a bed for the purpose of holding moisture and making nutrient elements available to the trees. Most of the time a manganese need should be supplied via a foliar spray.

Manzate and Maneb are both chelated substances with manganese sulfate. As with many fabricated products, there is inherent danger when the user is inexperienced. When chelates of this type are used in a blend of things, and the soil is high in calcium, too much can cause all the leaves to fall off the trees. There is no such danger with manganese sulfate.

Shakespeare once wrote, "There are more things in heaven and earth, Horatio, than are dreamt of in your philosophy." Hamlet was speaking about spirits, but he might as well have had modern agriculture in mind. We can hint and we can get specific, but we must never lose sight of the fact that we know so little. And we must also thank God that Carey Reams passed on to us a body of knowledge that might have died for generations had he not been available to bridge the gap. Our charge is, and remains, to improve upon what we have been gifted.

14

COLLOIDAL CLAY AND SAND

During the 1930s, when Carey Reams routinely journeyed to the University of Missouri for his Socrates sessions with the head of the Department of Soils, William A. Albrecht, the "Little Professor," was well into his classic colloidal clay experiments. Students at the university during those years remember the Reams visits and the likely topics up for discussion, yet we are forced to speculate what was said, unless we turn to a precise statement of the objective and the method used which was published as Agricultural Experiment Station Research Bulletin 60, *The Chemical Nature of a Colloidal Clay*, June 1923, Richard Bradfield, author, Drs. M.F. Miller, W.A. Albrecht and F.L. Duley (of University of Missouri) and Drs. F.E. Bear, Edward Mack and E. N.Transeau (Ohio State University), godfathers. Additional work on the subject by one of Bradfield's students, Leonard D. Bauer, has passed into the literature as Missouri Agricultural Experiment Station Research Bulletin 129, *The Effect of the Amount and Nature of Ex-*

changeable Cations on the Structure of Colloidal Clay, 1929. Indeed, there were many papers, all highly technical. Albrecht described his own role in these studies as follows: "I separated the finest part of clay out of Putnam silt loam by churning in a centrifuge running at 32,000 rpm after the clay had been suspended and settled for three weeks. At the bottom, that clay finally plugged up the machinery. But we had thinner and thinner and smaller and smaller clay until about halfway up in that centrifuge, there we had it as clear as vaseline. We took the upper half of that clay. We made pounds and pounds of it. We put it into an electrical field and made it acidic. We took off all the cations so it was acid clay. That was how we studied plant nutritions. We put on different elements in different orders. We mixed them, balanced them." Albrecht knew he had to begin with calcium. Extensive research projects served up this working code for balanced plant nutrition: hydrogen, 10%; calcium, 60-75%; magnesium, 10-20%; in some plants, 7-15%; potassium, 2-5%; sodium, 0.5-5.0%; and other cations, 5%. "While the above ratios are guidelines," wrote Albrecht in *Soil Reaction (pH) and Balanced Plant Nutrition*, 1967, "they have been found most helpful for humid soil treatments as more nearly balanced plant nutrition for legumes. They are also a sound reasoning basis for better growth of non-legumes. Those same ratios between the nutrients, emphasizing calcium almost ten times higher, and more, than others among the five, should make us believe that one is apt to find the calcium the more commonly deficient nutrient element for crops, too long covered in our belief that soil acidity calls for a cheap carbonate, being taken as limestone." The nutrient code expressed above is being used by the important laboratories serving eco-agriculture. The logic of Albrecht's search caused him to look at sulfur and the trace elements (which suggested themselves as reasons for the low protein values of crops being grown). The conclusions seemed obvious. "The plant's struggle is one for its synthesis of proteins," Albrecht summarized in *Soil Reaction (pH)*, "its living tissues, giving growth, self-protection and reproduction. Humans and all other warm blooded bodies are struggling similarly for adequate proteins. Those are synthesized from the elements as the starting point of creation by only plants and microbes; hence the soil, its microbes (centered about soil organic matter) and its crops, are the quality control of man's nutrition and of all the animals supporting him. In our management of the soil we are not yet the equal of what nature was before man's advent on the scene to take over that responsible part of creation."

By titrating nutrients to the naked clay, Albrecht uncovered what nutrient shortages and nutrients completed do to crops. Wheeler McMillen, the long-time editor of the *Farm Journal*, summarized in *Harvest, An Anthology of Farm Writing*:

> Dr. Albrecht, as a director of all of the Sanborn Field research, had seen these same acres produce face-reddening facts. For instance, he knew that back when everyone was talking and preaching crop rotations, evidence from Sanborn Field had proved that such practices under certain conditions could be not beneficial but actually very harmful. This man of classroom and laboratory, a born teacher, knew, too, that contrary to early admonitions that legumes "left the soil better than they found it," wasn't always true. Legumes, overdone, could — instead of leaving the soil with an abundance of stored nitrogen — leave it as impoverished as a share-cropper's land following a life-time of following a "one-crop system." Sanborn Field also had taught the research men that fertilizers applied "without rhyme or reason" could be almost as bad as no fertilizer at all.
>
> Albrecht knew these things, because he knew Sanborn Field, those few little acres that the unimaginative had once tried to turn into a parking lot, but which were capable of producing more scientific surprises than a clown in a circus. It was this knowledge which mothered a remark by him when the soil sample that had produced the then newest of the wonder drugs was presented in special ceremonies at the Smithsonian Institute.
>
> He declared at that time, his friends say, that it was very doubtful if any other plot of land, comparable in size on the face of the earth, has produced as much genuine knowledge and wisdom for humanity's use in combating physical suffering and hunger "as have the hallowed acres of Sanborn Field."

These lessons were not lost on Reams. His native Florida inspired a memorable response to the question, *And what do you think about soil in Florida?* The answer: *It would be a good place for some!* Many of the acres he engineered were little better than Albrecht's naked colloidal clay. They were white sand, and as an agricultural engineer he had literally to titrate on the needed nutrients in the right balanced. During most of his life he was guided by the Morgan Universal Testing Systems and the LaMotte procedures. The first thing he did was apply approximately one ton of soft rock phosphate. In those days the cost was $5 to $10 a ton. The next thing he did was apply high calcium lime, and then he usually laced the fields with several tons of cage layer chicken manure, not broiler litter. To set this complex assortment of soil nutrients and microbial food in motion, he added 200 pounds of ammonium sulfate per acre. That was his plan, to build a base magnetic field over the soil and to enhance life in the soil. In other areas of the country, soils are much better than blowsand, and therefore the fertility management program of necessity has to be different.

I have worked with some soils in southern Mississippi and southern Alabama that have the status of poverty soils, and they produce poverty crops. The basic principles discussed here apply in all these cases, if the farmer has the money to do what is indicated. There are several systems and conceptualizations for getting a fertility program off ground zero. The LaMotte test is one I favor. Commercial laboratories have their procedures which, too, are helpful in terms of the premises used.

I do have a specific procedure for row crops and seed crops. Sampling, of course, rates attention first, last and always.

When soil samples are taken, samples must cover different soil types through the sampled area. Fertilization for specific soil types portends a dilemma. It is possible to have quite a variation in soil types within a field. I have in mind a 160 acre field in Iowa. Well crafted soil maps indicated two very distinct soil types within that field. We were able to develop a work place to handle each type separately, using two specific programs.

Soil samples should be taken to a six inch depth. It is also imperative to put a date, a time and some other identification on the field — and if sent to a laboratory, a name and telephone or fax number.

Other inventories of facts must be assembled — the proposed crop and the field's history, namely crops raised on that particular piece of geography in the past as well as previous fertilization programs.

A history of previous fertilization practices is needed if a withdrawal is to be managed correctly. For instance, if high levels of anhydrous or potassium chloride have been used, it is never possible to back away as much as desired. My experience has been that if 180 to 200 pounds of anhydrous have been used per acre, the acre will likely end up with 130 pounds of some form of anhydrous that first year. Theoretically, if the farmer has been getting 140 to 150 bushels in the past with an anhydrous program, there should come a day when he will do the same thing with 80 pounds of anhydrous on a consistent basis safely. Some very pragmatic considerations figure in structuring any program. How much money is the farmer accustomed to spending for a crop? Equally important, how much money do neighboring farmers spend? If the farmer is unable to say how much has been spent in the past, the consultant is treading on thin ice, and the farmer is looking for a miracle. I swear that I can take two farmers, give one the best plan in the world, and he will make it fail. I can give the worst plan to a second farmer, and he will make it succeed.

When planning a fertility program, the time of the year this planning takes place is important. The fact that the farmer knows something about previous yields and input costs automatically provides an indication of rainfall in an area. This is where you plug in the many special observations covered in this book. Do you really understand carbon? What and how does it control? Air circulation and the humus level figure in the equation. If the calcium is low are the sulfates also real low?

The age old problem of acid-and-alkaline requires steady scrutiny, with full appreciation of what pH means and what it does not mean. If a soil is tight and permits no circulation of air, it will be probably both acid and alkaline. If you were to run a water soluble test on this, more than likely you would find no calcium, but this would suggest a fair amount of calcium but no energy. There is a requirement for carbon and air circulation.

If sodium is really high, say, 1,000 in a certain instance, this means that a serious use of gypsum may be necessary, perhaps

THE MORGAN TEST

The name M.F. Morgan is known to all workers in soil testing. Morgan was not only one of the earliest workers in this field but he was also a pioneer in pushing back the frontiers of science on this phase of soils research. His *Universal Soil Testing System* is a monument to his researches.

In 1927 he started research on soil testing with the designing of a porcelain soil test block for determining soil reaction in the field by means of indicators. The second step was the development of a nitrate nitrogen test with diphenylamine, using the soil reaction test block. Then followed the addition of tests for ammonia, phosphorus, calcium, aluminum, chlorides, and sulfate, using separate extractants for each, and the introduction of the artist slab or spot plate. The original test block was retained for reaction tests but was enlarged to permit three tests to be made simultaneously.

In 1935 Morgan introduced the highly buffered Universal extracting solution, which permitted all of the principal tests to be conducted on portions of one extract, and substituted the filter funnel for the Morgan test block. At the same time the following tests were added: potassium, magnesium, manganese, iron, nitrite nitrogen and sodium. The pH meter was adopted for reaction tests in the laboratory but the soil test blocks were still used in the field.

By this time the Morgan System had become widely known and adopted in many other states and foreign countries. Subsequent development took the form of added tests — boron, zinc, copper, mercury, lead and arsenic.

Morgan's last soil test bulletin, Number 450, entitled *Chemical Soil Diagnosis by the Universal Soil Testing System*, published by the Connecticut Agricultural Experiment Station, was unusually well received, and was soon exhausted, and reprinted. His work became incorporated in the LaMotte test Reams and Skow introduced to countless clients. — *The Connecticut Agricultural Experiment Station*

500 pounds per acre. Nitrates should be somewhere near 40 pounds per acre.

Gypsum is calcium sulfate. It has a tendency to act like baking soda, to fluff and drive the particles of the soil apart. Calcium carbonate does not do that.

Let's consider a soil with anaerobic bacteria quite high. Aluminum could flip-flop in such a situation, but probably remain low. The soil would be sour and highly alkaline — with lots of calcium unable to release its energy due to a lack of air flow, carbon and water circulation.

15

A FARM TESTAMENT

I have more than a few notes I would like to hand off as I approach the last chapter of this book. The crop, after all, is the thing and after planting there are the long weeks of waiting and monitoring. The idea that a crop is set for the season with planting is truly the benchmark of the amateur. So let's recap.

Tillage of the soil should proceed with aeration in mind. There are two things that fly 90 to 100% of the time — calcium sulfate and liquid calcium. There is a calcium on the market called Promasol. A liquid humate might also be used. RL-37 is a proprietary brand that almost always works if the others do not. In many soils, gypsum at 500 pounds per acre and liquid calcium at one to three gallons — perhaps up to six gallons per acre — all are indicated. There would seldom be a requirement for more than a gallon of the liquid humate per acre, and certainly no more than two quarts of RL-37. Most of the time one quart of the RL-37 would be adequate. It is a very good product when high chemical toxicity is reported. It brings down toxicity and loosens the soil. Aeration allows microbial life to do its thing.

There are several programs that can be plugged into the situation, all to good effect. If a soil has plenty of nutrients, simply doing what has been outlined above and adding 28% nitrogen—perhaps 20 to 25 gallons to the acre — will suffice, even under dire economic stress. If I were to add an expense, it might be for a wetting agent. However, I have found RL-37 does a fair job by itself. A good response to this program is a loose, spongy soil within 60 to 90 days. It may be modestly sticky, but the texture will be better.

The objective here is to patch a weak link in the chain–aeration, in this instance — which will release calcium. There is calcium hydroxide in this material. It works via the homeopathic principle to release calcium in the soil.

Ortho Chemical makes a 20-10-20. It is a dry material that contains trace minerals and services the soil well at 200 pounds to the acre. The reason I mention some of these materials is to argue the case for testing samples to see whether they can be utilized in the program. Another product that should be scrutinized is 12-0-0-26, Thio-Sul, three or four gallons to the acre.

It will be noted that I like to work with water solubles. Some of these products suggest application with a planter, not on the seed, but off to the side of the seed.

If planting is east to west, application should be to the north side. There may be some problems with this — namely being able to plant only one way. Regardless of what the crops are, the principles and procedures are practically identical even though the products may change.

Generally, I have found that as you bring the phosphorus-potassium ratio in line, there are fewer weeds, better aeration and more microbial life figures. The broad-spectrum remedies and the fine-tuning approaches both work, even though neither have made it into Weed Society thinking. This does not trouble me. I know of too much science that appears to be gobbledygook. And I think I might digress a moment to make a point.

Near Grand Prairie, Alberta in 1973, tree roots were fossilized in moments when a high-voltage line came down. The find was sent to the University of Regina, Saskatchewan for a KAr test. Many readers will identify the KAr, or potassium argon, test from its newspaper exposure along with the uranium lead (UPb) test and the more famous carbon-14 (C-14) dating procedure. KAr is often used to date the age of rock and soil samples. In the Alberta case, word came back that the test would be meaningless because it would date the fossilized roots as millions of years old because intense heat was involved in the petrifaction process.

KAr testing was used recently to date volcanic lava sampled on Hawaii's big island. The readout came back — three billion years. Subsequent checking revealed that the rock sample involved came from an eruption in the year 1801.

I offer these few notes as an aside because some few students— even after taking short courses — are worried about the Milhouse Unit conceptualization and about the Biological Theory of Ionization itself. Obviously we cannot offer more than the framework for our thinking and the results.

The results achieved in dealing with nut trees are what we can see and understand best of all. One of the biggest problems in maintaining nut trees is the failure to keep enough phosphate of manganese available to the tree. The best and cheapest way to supply these nutrients is via foliar spray.

Earlier in this book, much attention was given to the business of measurement, special attention being directed to the size of an acre. An acre is 43,560 square feet, and the square root for this figure is 208.71, meaning that an acre is 208 x 208 feet — figure it at 210 feet if you like. This will help you compute the number of rows per acre. Using these data, you can predetermine how much energy in a product is used up per acre of soil in a year, and how much energy it is capable of giving up if the yield is known. Let's take a corn crop that yields 5,600 pounds at 20% moisture. On a 100% dry pounds basis (8 x 5,600) this crop computes to 4,480 pounds.

The basic rule that 80% of crop nutrients come from the air must now be reviewed. To compute how much of the dry weight actually came from the soil, deduct 80% of the crop nutrients that came from the air (4,480 -3,584 = 896) to get the 896 pound figure. Of that 896 pounds, 80% of the energy it took to grow came from calcium. This means 179 pounds of the dry weight came from other sources. Of course we have not considered the roots and stalks. In theory, even though we take off 896 pounds per acre, in fact the stover and root system that came from the air remains in the soil. In short, we should have a net gain at the end of the year instead of a loss. And yet a lot of farmers are having a net loss. The difference between 896 and 179 is 717 pounds.

Here is a rule that must be considered and remembered. The value we computed at the beginning of the year had to be equivalent to double the one stated because the soil can't give up its contribution in one year. In order to grow a 100-bushel corn crop, at least 1,600 pounds of calcium must be in the inventory and maintained during the growing season.

The calcium, however, is not lost completely when its energy is used. It can be recharged. But you reach a point where the calcium energy field gets lower and lower, and is therefore unable to maintain the rate of release. That is why the calcium level must be maintained at all times, and not only at planting time.

As I close this lesson, reminders become routine and needed. One part carbon will hold four parts water. The more carbon in the seed, the quicker it will sprout, and this boils down to having more sugar and minerals in the seed.

Remember, all elements in a molecular structure are the same size under the same temperature and pressure. The center core of an element tells whether it is an anion or a cation, and nature will follow the line of least resistance.

The greater the density of the soil without humus, the greater the specific gravity. The lesser the density of soil nutrients, the smaller the yield.

The process of osmosis is not limited by time until a plant starts to set seed, at which point time enters into the process. The less time it takes to grow a crop, the better the quality. The higher the sugar and mineral content of plants and trees, the lower the freezing point. Top quality produce will not rot, it will dehydrate.

All organics are cationic or positively charged.

Plants live off the loss of energy from the elements during the synchronization of these elements in the soil, and the one thing we do not want to happen is to completely synchronize. We want the elements to try. When Carey Reams was in college, he had a botany exam. The professor gave him a leaf, and he was to write a paper on that leaf. Everybody handed in two or three pages on that leaf, and the professor took one look at each one and tore it up. He said, *You are going to have to come back tomorrow and write it over*. The next day, they all came back and were presented another leaf. Some got a half page this time, some a little less and some had a lot more. They gave them to the professor. He looked, tore them up and threw them in the wastebasket. About the third day, the students were really puzzled. Finally, one of them wrote down about three words, the professor kept it. He got what he was looking for. All he wanted on that piece of paper was that every leaf is different.

Remember, every single thing is different and that is why every situation has to be handled differently. Just because a fertilizer reads 10-10-10 and another reads 10-10-10 does not mean they are the same.

Like things attract each other. For every cause there is an effect.

Phosphate controls the sugar content of a crop. The higher the phosphate content of the soil, the higher the sugar content of the part the farmer sells. The higher the sugar content, the higher the mineral content. The higher the mineral content, the greater the specific gravity of a given bushel, box, bale, or bin. The greater the specific gravity of the product, the healthier the plant or the animal.

All elements must go into the plant in phosphate form except nitrogen if a healthy cell is to be expected. The ratio of all crops — except grasses — for phosphate and potassium in the soil is two parts phosphate to one part potassium.

The problem is what kind of a test are you using to make this statement? There are so many ways of testing soils and so many interpretations that the only thing we do know is that when you have a tendency to have higher phosphate and lower potassium, there seems to be quite an explosion in yield, especially in grasses. This also seems to be true in corn.

A fellow in Illinois who I worked with did quite a study on all the high yielding test plots throughout his area, and there was one very distinct characteristic in nearly all situations. The highest yields were in areas with the highest phosphate levels.

Potash determines the caliber of the stalk and leaves, the size of the fruit and the set of the fruit. Potash keeps fruit from dropping off the trees and it also keeps oats and small grain crops from dropping seed.

Nitrogen is the major electrolyte in the soil. Only that plant food which is soluble in water is available to the plant. Cationic substances go down. Anionic substances go up. Root crops have a tendency to be more anionic than crops above the soil.

In the box in the next chapter I have listed some of the deficiencies. Remember, when you make a statement that a specific element causes a deficiency, there are probably other things that enter into the picture — organic matter, water, tie-ups or other elements, etc. It's not necessarily a specific element that is lacking in the soil. It might be there, but unable to get into the plant.

Some years ago, Dutch elm disease swept the country. The trick to dealing with the disease is making phosphate available to the tree and that basically boils down also to getting the sap to flow. If you can get those two jobs done you can probably correct the problem, but there probably isn't enough time if the disease is already beginning to show.

The final question seems to be, can we really be close in making our computations and effecting an influence on cause and effect in the scheme of things? The crops say we can — and do.

When Apollo II returned to Spaceship Earth with moon rocks and dust samples, the uranium lead, or UPb test was invoked. The test produced four different datings for age of the rock:
- 4.6 billion years.
- 5.4 billion years.
- 4.8 billion years.
- 8.2 billion years.

Which figure is correct, or is any figure correct? According to the argon tests performed on the same lunar stones, the answer is none of them. The potassium argon (KAr) test says 2.3 billion years, according to *Science* magazine.

I do not know whether we will ever turn up the origin of the Milhouse Units Reams taught us as a conceptualization for God's plan, and not as a roll of the dice. It argues against the chaos theory, and for rhyme and reason. At least we can compute with relative accuracy the role of energy in agriculture, and our approximations are consistent, not wild guesses such as those we see with KAr, UPb, and C-14 "science."

Phil Callahan, an entomologist, has explained low-level energy as the force to move mountains in *Ancient Mysteries, Modern Visions*. And his theories dovetail with my own. The question of a role for chemicals of organic synthesis always returns, like a bad penny. We do not think these mutant chemicals have much of a place, and we think we have constructed a farm technology capable of operating without man-made poisons. Perhaps one of the insects studied by low-level energy expert Callahan will illustrate the point. Over the years there have been thousands of experiments in the laboratory that affect mutations. For at least 1,500 generations, the little fly has withstood the torture test to produce mutant offspring. In the process, science has produced shriveled wings, crooked bodies, weak eyes, no eyes and sterile flies. Not for nothing did one scientist remark, "Trying to improve an organism by mutations is like trying to improve a Swiss watch by dropping it and bending one of its wheels. Improving life by random mutation has a probability of zero."

We now know that radiomimetic chemicals (chemicals that ape the character of radiation), radiation itself, and many of the rescue chemicals used in agriculture can injure the chromosomes in the cell–plant, animal or man–either by altering the chemistry of a single gene so that the genes convey improper information (called point mutation) or by breaking the chromosome (called deletion). The cell may be killed or continue to live, often reproducing the induced error. Some types of cell damage cause genetic misinformation, that leads to uncontrolled cellular growth — cancer.

Knowing this much, we also know that a new mainline agriculture must gather speed and sweep ahead.

16

FOLIAR FERTILIZATION

Two concepts that tower above so-called "conventional" agriculture are now secure components of mainline farming for Century 21. Use of the refractometer for crop evaluation during the growing season has been well publicized for over a decade, and the metes and bounds of the technology and instrumentation are suitably covered in an earlier chapter. The second component — one equal to sound fertility management in the first place — is foliar fertilization.

Valid technologies have been used to make foliar feeds and liquid fertilizers since 1951, according to an *An Acres U.S.A. Primer* recap on the subject. {These developments [quote the *Primer*] went on virtually without the agricultural community knowing anything about them. Sylvan H. Wittwer, under whose leadership Michigan State University conducted investigations on foliar fertilization of plants, correctly summarized: 'There is probably no area in agricultural crop production of more current interest, and more contradictory data, claims and opinions, or where the farmer in practice has moved so far ahead of scientific research.' Even today the average land grant college has few, if any, scientists who have a working knowledge of using liquid fertilizers in growing crops. There will be those who object to this statement, citing Hanway's Iowa State University research, which came some twelve years after Wittwer's findings, but this hardly qualified because it failed as a result of using the wrong type of ingredients.

Credentialed and uncredentialed outsiders standing in the wings knew that Hanway's foliar N, P and K efforts would fail because the compound being used failed to answer plant requirements, would burn crops and also because it would cost too much. It involved the application of gross amounts of nitrogen, phosphorus and potassium to soybeans in three applications, using amounts required by holy writ of soil fertilization. As Sir Albert Howard would have pointed out, fertilization of such crops entailed a knowledge of many other things, such as what is a plant's requirement at each point in the growth cycle?

Working with nursery crops, flowers for market, strawberries, apples, pears and cherries, agronomists started answering these growth cycle requirements with seaweed extracts many years ago, all with successes that were spectacular, erratic, and rejected. There were things that could be done with the leaf that proved next to impossible when working with complexed soils.

Some few farmers knew this, but they didn't have a handle on the 'why.' In the early 1950s, Wittwer was working under arrangements with the Division of Biology and Medicine, Atomic Energy Commission. Experiments entailed the use of radioactive isotopes in assessing the efficiency of foliar applied nutrients compared to soil applications of those same nutrients. It was found that the efficiency of the foliar fertilizers was from 100 to 900% greater than the dry applied fertilizer materials. The results became a matter of record under Contract AT (11-1)-888, 1969, and soon touched base in select circles as an audio film entitled *The Non-Root Feeding of Plants*.

Some parts of that summary are of maximum interest. It was noted, for example, that deficiency disorders were becoming more common, that there were soil imposed problems of dilution, penetration and fixation. Foliar fertilization circumvented many of these problems. 'A plant's entire requirement for many micro-nutrients may often be supplied by above ground plant parts because the quantities needed are small and tolerances for the applied materials and rates of uptake are adequate,' Wittwer wrote. 'But for the macronutrients used in large quantities by plants and for most crops, only a part of the nutrient needs are satisfied, but the contribution can still be significant.'

Without using atomic tracer studies, T. L. Senn of Clemson University arrived at essentially the same conclusions, working both before and after the Wittwer studies. These studies have not been well-received in the main because they involved fertilizers inherently of medium and low analysis according to N, P and K folklore. Since it has been technically undesirable to distort these products so they contain an aggregate of 16, 20 or 24% N, P and K, they have frequently been denied registration."

As noted in the earlier chapters of this book, the Biological Theory of Ionization had cemented into place its own premises long before the above synopsis was made a matter of record. A more mature concept of how plants grow and what nutrients have to do with the part the farmer sells suggested a monitoring task week after week during the growing season, and repair work as indicated via the foliar route.

In developing a foliar program, maximum attention must be given to the thickness of leaves, how well leaves stand up, the degree of wilt, and so on. A thin or weak leaf suggests a nutrient deficiency, or low TDN — total digestive nutrients. The caliber of the stalk and stem is extremely important, as is the development of the root system. Field observation will reveal an under-developed root system when herbicides are used. These shortfalls can be repaired with foliar sprays and fertilization through irrigation systems.

In reading what a field has to say, color is of maximum importance, and can range from one extreme to another. According to problems encountered, plant color can go from blue-greens to pale yellows, a bright green being preferred with a black blue-green being the harbinger of trouble. Painting a field nitrogen green simply results in an excess of water in the plant.

The bottom line still boils down to identifying the basics. Front burner stuff is a check on the ergs as outlined earlier. If the ergs are low, this shortfall must be repaired if yields are to be maintained. The foliar route won't increase the ergs of energy very much because it takes a volume of plant foods to handle that chore. Usually side dressing and nutrient injection via irrigation systems are indicated.

There is an exception to the above. When there are a lot of reserves in the soil, the foliar route can sometimes change the way organic acids are excreted at the root level. A stronger organic acid excreted by the rootlet can better extract nutrients from the soil if, indeed, those nutrients are there in the first place. This situation is true when the soil system is low in humus, and it is usually very temporary.

When an ergs test is made, it is necessary to flesh out the information gathering synopsis with a refractometer reading on the plant. Weeds should also be checked with a refractometer to see whether the crop plant or the weed has the highest reading.

Raw data are of little value if they cannot be plugged in to the process called photosynthesis. Sometimes we get too used to hunting down esoteric texts and college tomes that obfuscate more than they enlighten. One of the best books I have encountered — for the purpose of making the subject come clear — has been my son's sophomore science book. I think it is the first book I've encountered that stressed the importance of potassium to the photosynthesis process. Potassium affects the guard cells around the stomata, how they are opened and closed.

Parenthetically, it must be observed that sound definitely affects the opening and closing of the stomata, as do temperature and moisture. A vibratory frequency near that of the plant will definitely help regulate the stomata orifice.

The key here is that when sugars are down, something is usually wrong with phosphate availability. Generally speaking, it is usually a case of a short base amount in the soil. Phosphorus may be there, but presence is not enough. It still takes a critical level to construct availability that is high enough to produce the sugars required to build a plant.

We can have quite an impact with foliars, but if the base amount in the soil is too low, it becomes difficult if not impossible to keep phosphorus high enough during the filling stage of a crop. That is why it is always best to build both phosphate and calcium levels with non-acid treated resources while rootbed preparation is underway.

In the foliar application, the prime and best quality source of phosphate is food-grade 85% phosphoric acid—or good white phosphoric acid.

To build a foliar spray, the above element comes first — then water. The amount of moisture in the atmosphere rates maximum attention. If the air is dry, the low end of the recommended amounts should be used to construct the spray. Humid territory suggests a higher level of nutrients in solution. This translates to using half a pint to a pint of phosphoric acid per acre when humidity is high, and less than half a pint under dry conditions. Using a conventional sprayer, usually 20 gallons of water to the acre is correct. A mist blower — such as a Chiron sprayer — would work best with a pint of phosphoric acid in 100 gallons of water.

In addition to phosphoric acid, some source of nitrogen is indicated. One of the best is household ammonia. In some areas of the country this is available under the Bo-Peep brand name. Here, again, a pint to the acre must be suggested, the usual variation for dry time and wet season being considered. The mix can be higher in wet situations, and lower in dry situations, the key consideration being not to harm the leaves. Once the phosphate and nitrogen situati — has been addressed, other sources of nitrogen can be considered—liquid fish, for instance. Fish usually runs about 5% nitrogen and a half a percent phosphate, 1 or 2% potassium being the norm.

Seaweed foliars also work quite well. But from my chair, they do not deliver quite as well as liquid fish. It may be that the potassium shortfall underwrites my conclusion.

If foliar application is to be made during a seed setting stage of crop production, ammonia — a cation fertility input — will help set those seeds and fruit, and start pointing to potassium, whether you need it or not. There are several sources for potassium, and anywhere from one-fourth pound to two pounds potassium sulfate per acre should be considered. Again, weather conditions and severity of the shortage will determine the formula.

Filtered soft rock phosphate water is often of value in formulating a foliar spray. Such a water helps the spray stick to the leaf.

If analysis reveals a shortage of iron, iron sulfate can be introduced into the solution — a pint to two quarts of iron sulfate dissolved in the water suggesting a norm.

The foliar spray can be a veritable cafeteria of plant nutrition, each pick and choose entry answering a requirement uncovered while the crop grows. The basics mentioned above can be supplemented by things such as vitamin B. These work as growth regulators and merely suggest the possibilities and the sophistication that can be achieved.

The character of the water carrier is also important. This means pH should be adjusted to between 6 and 8. Baking soda will raise the pH, and vinegar will drop it to the desired level. In any case, a pint to a quart of vinegar per acre will help put the spray on the frequency of the plant, and this will help achieve greater uptake. In most farm situations, well water is best. It must be checked, of course, because the potential for herbicide and insecticide contamination is worsening year after year. Spring water is inherently dangerous, especially when herbicides have been used in the general vicinity. Temperature of the carrier vicinity does not seem to matter. By the time the spray travels to the leaf, it will have reached ambient air temperature anyway.

There is nothing magic or occult about foliar treatment. To be effective, it has to answer a requirement, and this requirement can best be determined with a refractometer reading for the crop. My suggestion is simple in the extreme. Take a small starch spray bottle and treat an area in the field. Then come back in 20 to 25 minutes for a second refractometer reading. The brix scale will tell you whether the test spray was needed and whether it made a difference. A little testing and a little experimentation will uncover steps to be taken. If a brix reading moves up between a half point and two points in twenty minutes, the spray is delivering and the plant is enjoying the unexpected lunch.

How a foliar spray performs is governed by the calcium reserve in the soil. A low calcium soil will set up a resistance against the foliar treatment, the effect being either no response or a very temporary response. Even so, it is often possible to bump a crop just enough to move plant growth past a critical level. Even when treatment fails to improve the refractometer reading, and yet does not drop it, continuation of the foliar approach suggests itself. Small amounts of phosphoric acid, fish, seaweed extracts, household ammonia, as a combination, provide the basics for homemade foliars, the phosphoric acid being food-grade 85%.

I almost always try to put two quarts of fish and four ounces of powdered seaweed in my sprays. The mixes may vary, always answering the nuances discovered in the field. But basically I find that a fish and seaweed combination with phosphoric acid is quite effective in keeping insects out of the field, especially if the brix is below 12. This mixture will substitute for the sugars when the latter are not high enough for plant protection. Repeated sprays with fish and seaweed combinations in low amounts as a ten day program — especially in orchards — will gradually build up fruit-wood and root production for the following year. The consequences will be high quality produce. Apples will be firm and without blemishes. Moreover, they will exhibit good taste and flavor. B-12 added to sprays on a regular basis not only improves flavor, it also presides over improved brix readings. In working with fruit groves, it is mandatory to start a year ahead of time.

B-12, available as an injectable for animals, works quite well, 15 to 22cc per acre. The presence of B-12 in the foliar spray has something to do with chelation of calcium and making it more available. Indirectly it has an impact on fruit quality.

I have worked with one cherry grower who had significant improvement in his cherry crop after spraying B-12, albeit alone several times during the growing season.

Withal, foliar sprays are probably the best way to apply micronutrients, and they certainly represent the most economical approach. Effectiveness of these sprays is dependent on having a minimum of 2,000 pounds calcium in each acre of soil, basis the LaMotte system of testing.

One of the reasons academia dropped the ball on foliar spraying may be because the importance of phosphate in photosynthesis was overlooked. Carey Reams taught all his students, yours truly included, that it is pretty hard to beat quality phosphoric acid for dealing with the phosphorus-photosynthesis connection.

Manganese is a prime requirement for getting a good seed fill. This is especially true for stone fruit, peaches and apricots, for instance. Housewives who purchase grocery store fruit often encounter rotted centers, always a sign of manganese deficiency. Foliar application can prevent the problem.

Manganese sulfate will do, but the key is its mix with phosphoric acid. Application must be started a year ahead of time.

The same is true for nut trees. Yield can be increased significantly by spraying small amounts of phosphoric acid and manganese sulfate. Yield may also respond to liquid humates as foliars, which are now coming into the marketplace.

Liquid humates in foliars carry a caution sign. Levels have to be kept low, ever trending to the vanishing point. The liquid humate form is probably the most powerful chelating agent known to man. Thus the caution as to amount. Another caution for thoroughness of mix and blend has to be added.

We at International Ag Labs, Inc. have worked very hard at building a foliar spray that we think is very special. We have succeeded in making a solution mixing phosphoric acid and calcium together. Why is it important to mix these together? When we study what Dr. Reams taught about foliar sprays, he always wanted some phosphorus or soft rock phosphate in the spray. This was a problem for him, because the soft rock was not easily sprayed or handled. Also, the soft rock was not easily absorbed by

CAUSES AND EFFECTS

Boron Deficiency	Black heart potatoes
	Hollow stemmed alfalfa
	Hollow stemmed cabbage
	Generally, hollow stemmed means a boron deficiency
	Cabbage, oranges, apples split
Boron Excess	Strawberries are woody and hard
Calcium Excess	Scaly skinned vines
Manganese Deficiency	Peach pits split
	Deep eyed potatoes
	Pecans and nectarines split
Manganese Excess	Produce will go to seed before harvest
Cationic Nitrogen Excess	Produce will go to seed before harvest
Nitrogen Deficiency	Blue tint in the corn leaves
Nitrogen Excess	Bark slides off the root
Sul-Po-Mag Excess	Apples, oranges, cabbage split
Sulfur Excess	Decays at maturity
	Peaches never get ripe
	May not be enough lime
	Onions rot in the center
Potash Excess	Tips of leaves turn black
	Black spots on alfalfa
	Scaly skinned vines
Potash Deficiency	Watermelon vines

the leaves of the plant. The phosphoric acid was ideal for the absorption by plants, but didn't mix well with calcium.

When we study how Dr. Reams rated the different elements according to his biological theory of ionization, we have two complete opposites: one turning clockwise and one turning counterclockwise. This gives a tremendous energy release to the plant when these two elements are sprayed together. The problem has always been that when they were mixed, they formed tricalcium phosphate, which was not readily taken in by the plant.

I have seen many ingenious ways of handling this problem on farms. I have seen farmers with two tanks and two spray booms on the same tractor, one spraying 10-34-0 and one spraying liquid calcium. This got the elements on the leaves, but as soon as this combination hit the surface it was converted into tricalcium phosphate and, of course, again this hindered absorption. Let's look at some different ways of mixing these elements. I have seen many ways this has been done, but the first we will discuss is mixing phosphoric acid and liquid calcium.

This mix has about a 100% chance of turning to tricalcium phosphate. If you use pure enough products, it can be sprayed through a sprayer. This combination should have good agitation in the spray tank and should be compared to using wettable powders, but here again we have just made tricalcium phosphate.

The next process we will look at is one to keep the two elements apart in the solution. This is called an emulsion, whereby a chemical process has been added to the phosphoric acid and liquid calcium to keep them apart in the container. This can be identified one of two ways. If you look at the solution closely in a well lighted area, you will be able to see little waves floating in the solution. The other way an emulsion can be identified is to mix the ingredients and then let the solution sit undisturbed for a day. The elements will separate. The phosphoric acid (weighing 14 lbs. per gallon) will go to the bottom and the liquid calcium (weighing 12 lbs. per gallon) will rise to the top. This again needs agitation in the spray tank when sprayed, but not as much as using the first process. Although this process may be satisfactory, we do not have any information on what happens once the mixture hits the leaves. We tested this in our lab and decided against using it because there were too many unanswered questions as to what was happening on the leaf surface.

The last process we will look at is that of making a complete mixture of calcium and phosphorus without converting it to tricalcium phosphate. When we first started — or actually got close to making this go together — we had it made and were excited about the product, only to come in a few days later to find it had reverted back and separated out. This was a big disappointment, but we were too stubborn to quit after we had come so far.

We finally learned that not only the sequence that we used to combine the products in the tank, but also the the timing of that mixture, both were of critical importance. This meant that we were not giving the chemical process long enough to react in the solution before we added the next element. It also changed when we went from the lab to making large batches. The first large batch we made went out in the field, because it reverted back. We found the larger the batch, the longer it takes to make. So, at the present size batch we are making, it takes three days to manufacture one batch.

If you are familiar with our lab, you know by now that the product I refer to is our new label Amaze. We had been told by many people that this product couldn't be made. I want to thank our chemist Chuck Kirsch for his persistence when it would have been so easy to give up.

I have not been involved in chemistry to this extent during my lifetime, but what it gets down to is a chemical puzzle and all of the pieces must fit. When they don't fit, you can have the worst mess you have ever seen. One day, Chuck was working on one of his puzzles when he added a certain ingredient to his mixture and immediately realized it was wrong. He grabbed the flask and was trying to shake the contents out into the sink so he could save the beaker, but some of those mixtures get so hard you just have to destroy the glass.

I might add a little note as to how we named this product. When we had gone though all of these disappointing experiments in glue production and then finally came out with a product that was successful, the first word uttered was, "Amazing!" The word stuck to the new product like the glue stuck to the flask.

I believe that foliar feeding will be the new discovery by the fertilizer industry by the turn of the century. I was reading in a leading fertilizer trade magazine today that they thought trace minerals could be foliar fed but not macronutrients, so they are considering the practice. We have a lot of tests being done with foliar feeding from both a quantity and quality standpoint. We do not have the yield data in from corn and soybeans, but the alfalfa tests were running from 20 to 50 points higher in relative feed value and up to 3/4 ton per acre more yield.

The yield increase will only be found if you weigh the hay, it will not be seen by the naked eye. One farmer said he saw no difference until he hauled his round bales off the field and found that he had trouble lifting the bales that had been sprayed. Another farmer had weighed sections of windrows that had treatment and the control. He figured he got 800 lbs. more per acre on one crop of alfalfa.

We have been testing apples in Washington state that have been sprayed with the product. The tests came from three different orchards, one had yellow delicious apples and the other two were red delicious. The first thing we looked at was dry matter. An average of all three tested showed 11% more dry matter when sprayed with Amaze. If these apples are sold by weight, there is the yield difference. They also averaged 40% more calcium than the control, which makes a big difference in storability. The treated apples will store much longer since the higher calcium levels will make the cell walls much stonger in the apple.

Also, the treated apples averaged 10% higher in sugar. This, of course is the taste and ultimately is what sells apples. The Amaze-treated apples tasted much sweeter. Another thing we noticed was that the skin was tougher on the control apples.

The last factor found was that the protein was 20% higher. Apples are normally a fairly low protein food, but Amaze did raise protein. The rise in the protein content has been noticed in the alfalfa too.

To sum this all up, foliar sprays do work and are quite effective. Amaze is a 5-16-4-5 Ca foliar spray and is made specifically as a foliar.

The forage tests that follow on this and the next page are from the 1st and 2nd crops of alfalfa harvested off of the same field that had been foliar sprayed with Amaze. There were two applications made when each crop was two to four inches high.

ALFALFA RESULTS

Northrup King Multileaf, 1st Cutting

Test Performed	As Recieved	Dry Matter
Crude Protein	15.07	26.06
Calcium	1.14	1.97
Phosphorous	.20	.35
Magnesium	.15	.27
Acid Detergent Fiber	23.94	
Neutral Detergent Fiber	24.55	
Moisture	42.2	
Dry Matter %	57.8	
Salt	.18	.31
Digestible Dry Matter	70.25	
Dry Matter Intake	4.89	
Relative Energy Value	5,198.7	
Relative Feed Value	266.22	
pH	6.4	
Sugar	34	58

A & L MID WEST LABORATORIES, INC.

13611 "B" Street • Omaha, NE 68144 • (402) 334-7770 • FAX (402) 334-9121

REPORT NUMBER: 3-216-1030

REPORT DATE: 08/06/93

ACCOUNT NO: 9475

SAMPLES FROM:

SUBMITTED BY:

SEND TO:
INTERNATIONAL AG LABS INC
P.O. BOX 788
FAIRMONT, MN 56031-9802

Rev. 5.6-348-sef

FEED ANALYSIS REPORT

LAB NUMBER	SAMPLE IDENTIFICATION	MOISTURE %	CRUDE PROTEIN %	CRUDE FAT %	FIBER %	ASH %	TOTAL DIGESTIBLE NUTRIENTS %	NEl Mcal/lb.	NEm Mcal/lb.	NEg Mcal/lb.	DE Kcal/lb.	ME Kcal/lb.
22328	HAY	0.00	26.55		21.94		77.55	0.80	0.80	0.54		
		13.12	23.06		19.06		67.38	0.70	0.70	0.47		

LAB NUMBER	SULFUR %	PHOSPHORUS %	POTASSIUM %	MAGNESIUM %	CALCIUM %	SODIUM %	IRON PPM	MANGANESE PPM	COPPER PPM	ZINC PPM	SELENIUM PPM	COBALT PPM	MOLYBDENUM PPM
22328	0.24	0.29	1.34	0.28	2.29	0.039	105	48	23	29			
	0.21	0.25	1.16	0.24	1.99	0.034	91	42	20	25			

LAB NUMBER	NITRATE (NO_3) %	DIGESTIBLE PROTEIN %	ADF Pro %	pH	NDF %	RFV %
22328					24.80	269.0
					21.54	233.7

REMARKS

6) Fiber is an acid detergent fiber.

This report applies only to the sample(s) tested. Samples are retained a maximum of thirty days after testing.

A & L MID WEST LABORATORIES, INC.

On page 31 of this book appears a boxed entry on the conductivity meter. This instrument is almost as important to mainline farming, Century 21, as tillage equipment or a seed drill. Conductivity of the spray must be checked. This is equally important whether foliar application or irrigation delivery is used. When nutrients are added to irrigation waters, a reading of 2,000 on the conductivity meter is a prelude to achieving a good response. A reading in the 1,100 to 1,200 range suggests disappointment coming up. A reading with an equal spread above 2,000 portends a burned crop. Many of our findings are invading virgin territory. I know for a fact that when I get potato field ergs in the right range, production will increase by 30 to 40 bags per acre.

Related to the several procedures unveiled in this chapter are the dry water-soluble fertilizers, now reaching the support market. These are dry N, P and K materials bumped with biologics, sugars and other energy kickers. These are working out quite well, especially with tree crops.

My experiences suggest that if you kick dry water-solubles or even N, P and K formulations with fish and seaweed extracts, the resultant impact can be breath-taking, more so than if any one of the kickers is added singularly. These materials are quite compatible. By marrying fish and seaweed extracts, the combination can have a fantastic effect on brix readings.

When you deliver a higher level of dissolved solids into the plant, the result will be a healthier plant. This translates into better shelf life, improved storage capacity, and better flavor maintained over a longer period of time. Finally, in the field, this delivery of more dissolved solids into the plant confers immunity to bacterial, fungal and insect attack.

Many times an insect problem can be answered more effectively with a combination of fish, seaweed and phosphoric acid than by spraying pesticides.

The beauty of foliar technology is that it can be inexpensive technology. For instance, B-complex compounds can be made from live oak leaves soaked and fermented in a wooden barrel. That's a natural way of making B-complex compounds. The same thing can be done with Irish potatoes. The tuber, skin and pulp, can be put into a barrel of water for three or four weeks, depending on the time of year. In summer, the three- to four-week time frame should be appropriate. Three or four ounces of the resultant brew per acre usually is adequate.

There are other "natural" procedures. One system calls for the crop to be sprayed to be liquified. The liquid becomes a homeopathic remedy of sorts when sprayed two or three ounces per acre. This will set the spray on the frequency of the crop. Admittedly, some of this technology still needs refinement, but preliminary results say that the remedy is an appropriate substitute for insecticides and herbicides.

Obviously, the most valuable crops invite the most attention. For instance, alfalfa is an affordable target for foliar spraying, and spraying should be accomplished after every cutting, the sooner the better. Many times just a touch of calcium nitrate or potassium sulfate is very effective, especially when buffered with a miniscule amount of fish and seaweed. The foliar approach also suggests itself when the soil is quite compacted, since soil conditioners can be added at the same time.

As with many of the procedures I have discussed, foliars for alfalfa have a caveat in tow. A look under the crop canopy before cutting is always indicated. If weeds are coming in under the canopy, this means the energy level in the soil is too low. Now it becomes vitally important to apply a nutrient spray formulated for the alfalfa crop so that weeds will regress. A refractometer reading on the weeds and alfalfa should be juxtaposed for instant comparison. If the reading for the alfalfa is higher than the one for the weeds, there won't be much of a problem. But if the weeds have the highest brix reading, it means a shortfall of calcium in the soil. This situation suggests a foliar spray with calcium nitrate and use of a soil conditioner, or any of a number of biologics on the market. Nitro-Max suggests itself as a foliar post-cutting. The real secret is to make a foliar out of calcium and phosphorus, which is tricky. For one thing, they don't mix well and they have a tendency to react rapidly and become a material resembling cottage cheese.

The suggested remedy to offset this danger is to add sugar to phosphoric acid and mix, then add calcium nitrate. If the phosphorus is kept below 3%, calcium can be added to the mixture, the result being a phosphorus, calcium and sugar mixture that is extremely effective in building the energy field required by alfalfa, yet one hostile to weeds.

One last note must now be added to this lesson of foliar nutrition. Many farmers have moved over to the mist type sprayer. Under conditions of high humidity this type of equipment indeed presides over the best possible distribution on nutrient rich mist to the leaf. If extremely dry conditions prevail, a high evaporation loss becomes a danger. The mist may not make it to the leaf.

There are a few things that can counteract this problem. Phosphorus in the spray will help the mist make it to the leaf because it serves as an anchor. Carey Reams once gave me a formula he used to achieve rapid dry-down of alfalfa hay. It involved sea salt, vinegar, molasses and phosphoric acid:

10 gallons of seawater.

5 gallons of black strap molasses.

1 quart of household ammonia.

5 pounds of Calphos.

Add water to make 100 gallons of mix and use three gallons of mix per acre.

Withal, the foliar route must be considered the only valid rescue route. Chemicals of organic synthesis for the purpose of bacteria, fungus and insect control do not represent intelligent crop management. That role has passed to the diversity of nutrients discussed in this book, and to the innovators who manage them with intelligence that bespeaks a measure of insight into the Creator's plan.

INDEX

Leonardo of Pisa, 88
Anatomy Life and Energy in Agriculture, The, 6
light, speed of, 24
light, 130
Light, Radiation, and You, 161
lipids, 3
liquid fish, 115
liquid calcium, 131
Lorentz, 46
lye, 55

Mack, Edward, 169
magnesium, 12, 13, 63, 66, 112
magnesium sulfate, 113, 139
magnetic flow, 18, 22, 24
magnetic field, 22, 46
magnetic pull, 24
magnetic charge, corn stover, 27
magnetic flow, 89
magnetic susceptibility, 160, 161
magnetism in soil, 25, 26
magnetism, 33, 67, 80
Maneb, 167
manganese sulfate, 168
manganese, 47, 48, 59, 60, 63, 167, 191
manganese deficiency, 198
manganese excess, 198
manure, chicken, 10
manures, 12, 151, 152
Manzate, 167
mastitis, 9
McMillen, Wheeler, 171
Mendeleyeff table, 40
mercury, 163
metaphysics, 37
Mexico City, latitude of, 24
Michigan State University, 185
microbes, in soil, 27
microorganisms, 9
Milhouse Unit, 29, 30, 40, 42, 96, 120, 121, 179, 183
Miller, M.F., 169
milorganite, 152
mini-magnetic field, 23
Minkowski, 46
mitochondria, 12, 14
Mock [Einstein's teacher], 46
monammonium phosphate, 9
monocalcium phosphate, 135
Monsanto Chemical Company, 132, 133
moon, 124
moon phase, 39
Morgan Universal Testing Systems, 172
Morgan test, the, 174
muriate of potash, 10

muriated potash, 115

N, P & K, 46, 53, 73, 75, 90, 91, 135, 144, 199
N, P and K fertilization, 3, 9
Nagasaki, 46
National Geographic, 5
New Testament, 131
Nile River, 18
Nile flood, 43
nitrate nitrogen, 9, 13, 47, 77, 134, 157, 161
nitrate, 63
nitrate of soda, 50
Nitro-Max, 200
nitrogen, crop requirement, 28
nitrogen, 2, 3, 9, 34, 35, 47, 48, 65, 69, 73, 76, 90, 111, 144, 157, 161, 166
nitrogen deficiency, 198
nitrogen, excess, 65, 198
nitrogen, nitrate, 9
Nitrous psedomonas, 80
No-till, 69
Noah, 23, 40
Non-Root Feeding of Plants, 185
Noordenstrong, Bjorn, 159
north-south rows, 23
North Temperate Zone, 23
North Pole Magnetic stream, 23
Nukus (Siberia), 5
nutrition, 162

oats, 7
octocosanols, 3
oils, 158
organ, 158
organic matter system, 91
organic matter, 70, 71, 140
organic, 67
organic materials, 72
organic acids, root, 188
organically soluble calcium, 9
organics, natural, 50
organics, 181
organism, 1
ORP (oxidation reduction potential), 146
Ortho Chemical, 179
osmosis, 84, 104, 181
osmosis, 104
Ott, John, 159, 161
oxidation, 76
oxygen, 6, 42, 76

paramagnetic soil, 161
paramagnetic stone, 18, 25, 43
Paraquat, 57
peat, 72

ACKNOWLEDGMENTS

Most of those who contributed to the crafting of this book have been acknowledged, amply we hope, in the body of the book itself. Still, there are a number of people who deserve to be mentioned by name. First is the late Dr. Carey Reams, whose lectures attracted Dr. Dan Skow and Charles Walters Jr. over a decade ago. Some of his taped messages have been transcribed and published in *Acres U.S.A.* Dan Skow's classroom fare also helped background this text. Special thanks go to Wendell Owens, a Skow Enterprises staffer, who exhibited great tolerance and fortitude in checking and rechecking the finished product; to *Acres U.S.A.* staffers Brad Frisch and Shirley Renicker, who performed above and beyond the call of duty in bringing the manuscript to print; to Paul Cupp, who provided art illustrations; to Phil Callahan, who read the manuscript critically and offered improving suggestions; to Katie Rosenberg, for her cover portrait shot of Dr. Skow, and to Transnational Agronomy for the cover background.

Writers and agronomists do not live by words and soil systems alone. They, too, need compassion and gentleness, both of which have been provided by our colleagues. We thank them all warmly.